CELL SURFACE RECEPTORS:

A Short Course on Theory and Methods

Third Edition

CELL SURFACE RECEPTORS:

A Short Course on Theory & Methods

Third Edition

by

Lee E. Limbird, Ph.D.

Professor of Pharmacology

Vanderbilt University Medical Center, Nashville, TN

Library of Congress Cataloging-in-Publication Data

Limbird, Lee E.
Cell surface repectors : a short course on theory & methods / by Lee E. Limbird. – 3rd ed.
p. cm.
Includes bibliographical references and index.
ISBN 0-387-23069-6 (alk. paper) – E-book ISBN 0-387-23080-7
1. Cell receptors. 2. Binding sites (Biochemistry) I. Title.

QH603.C43L56 2004
615'.7—dc22

2004058276

Printed in the United States of America.

9 8 7 6 5 4 3 2 1 SPIN 11055082

springeronline.com

Research is essentially a dialogue with Nature. The important thing is not to wonder about Nature's answer–for she is always honest–but to closely examine your question to her.

A. Szent-Györgi, a paraphrase

CONTENTS

PREFACE

In this, the third edition of *Cell Surface Receptors: A Short Course on Theory and Methods*, I have tried to link theoretical insights into drug-receptor interactions described in mathematical models with the experimental strategies to characterize the biological receptor of interest. I continue to need to express my indebtedness to my earlier tutelage in these areas by Pierre DeMeyts and Andre DeLean, which occurred during my postdoctoral years as a member of Robert J. Lefkowitz's laboratory at Duke University. Other concepts, particularly classical approaches to defining and characterizing receptors, I learned from Joel G. Hardman while teaching a course together at Vanderbilt School of Medicine on receptor theory and signal transduction mechanisms. My national colleagues also have been terrific teachers, including Terry Kenakin (Glaxo Smith Kline), Harvey Motulsky (GraphPad Software, Inc.) and Rick Neubig (University of Michigan). In the end, of course, the motivation of preparing such a text is for the students, whose contagious enthusiasm encourages efforts to meet their needs. I hope this text is of value to investigators–at whatever stage of their career they find themselves–who want to identify, characterize and understand the biology of a receptor of interest.

I prepared this revision just prior to taking a sabbatical from Vanderbilt University. Vanderbilt has been a wonderfully supportive and intellectually stimulating place to work and to continue to learn. I am grateful to Eric Woodiwiss, for his technical support in preparing the manuscript and the figures, and to Harold Olivey, Ph.D., a former student in my courses at Vanderbilt, who read and thoughtfully critiqued the text. Without their help, I suspect this edition would not have materialized from draft to completion.

The study of receptors has changed considerably over the period of the publication of the three editions of this book. The cloning of several genomes makes it unlikely that preparations of receptors now or in the future will arise from their purification as trace proteins from native tissues, but rather from a myriad of molecular approaches. Nonetheless, understanding the molecular mechanisms and ultimately the *in vivo* biology of these receptors means that investigators will engage in molecular, cellular and ultimate in vivo strategies. To work across this continuum means that we must be forever grateful to the remarkable insights of those early describers of receptor theory and the criteria expected for biologically relevant receptors. We are the beneficiaries of their genius, simply fleshing out a skeleton, a conceptual framework, that preceded us by decades.

Lee E. Limbird
Nashville,
Tennessee

A good question is never answered. It is not a bolt to be tightened into place but a seed to be planted and bear more seed toward the hope of greening the landscape of idea. ***John Ciardi***

1. INTRODUCTION TO RECEPTOR THEORY

Much of the conceptual framework regarding how to study receptor function evolved from pharmacological investigation of drug action. Consequently, the historical account of the development of receptor theory in this chapter will emphasize early investigations of drug action rather than (for example) physiological studies of hormone action. However, the reader must keep in mind that the term drug can be defined as any chemical agent that affects living processes. Drugs bind to receptors presumably designed for interaction with endogenous hormones and neurotransmitters or other regulatory agents. **Agonist** drugs are analogous to endogenous hormones and neurotransmitters in the sense that they elicit a biological effect, although the effect elicited may be stimulatory or inhibitory. Different agonists activate receptors along a continuum of effectiveness; those which induce or stabilize less productive conformations are termed **partial agonists**, a property which will be discussed in considerable detail later in this chapter. In contrast, **antagonist** drugs are defined as agents that block receptor-mediated effects elicited by hormones, neurotransmitters, or agonist drugs by competing for receptor occupancy. Antagonists, as initially defined, were competitive inhibitors of receptor occupancy by agonists, having no intrinsic activity in their own right. However, more recently, antagonist agents have been observed to have negative intrinsic activity, or behave as **inverse agonists**, and decrease "basal" (agonist-independent, or constitutive) receptor activity. Still other antagonists of function mediate their effects by interacting with another, *allosteric*, site rather than in the binding pocket of the native agonist (defined as the *orthosteric* site) (Christopoulos and Kenakin [2002]; Kenakin [2004]; Neubig et al. [2003]). The properties of agents that interact via the orthosteric binding sites of the receptors are shown schematically in figure 1-1.

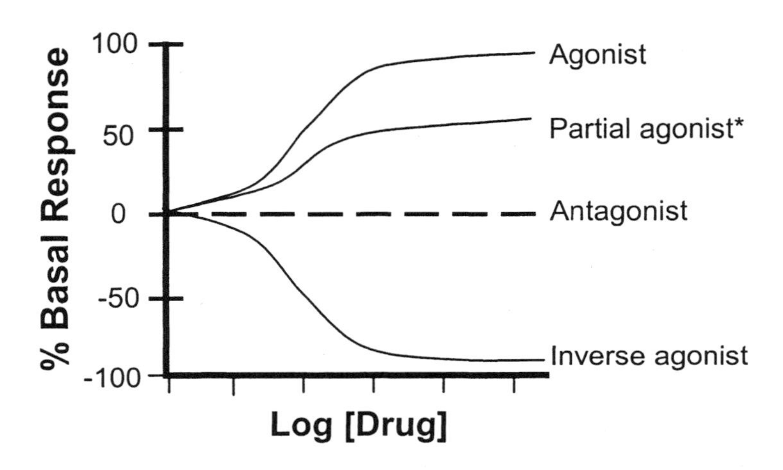

Figure 1-1. **Schematic representation of the functional consequences of the binding of drugs at the site of binding of endogenous ligands (the orthosteric site).** Agents which activate the receptor are *agonists*, and can elicit fully efficacious or partially efficacious (*partial agonist*) properties. *Partial agonists can either elicit a full response, but with lower efficiency or efficacy than full agonists, or, as shown in this schematic, elicit a submaximal response compared to a full agonist, even when fully occupying the receptor population. The properties of partial agonists and the theories that describe their behavior are considered in detail in later sections of the chapter. Classic, or null, *antagonists* occupy the agonist binding pocket and block receptor-mediated function by blocking agonist occupancy and subsequent agonist-elicited responses. *Inverse agonists*, or negative antagonists, stabilize inactive receptor conformations and decrease "basal" receptor activation in a dose-dependent manner.

ORIGIN OF THE RECEPTOR CONCEPT

Contemporary scientists take it as a "given" that biological substances such as hormones and drugs elicit their effects via interaction with specific *receptors* in a manner analogous to the interaction of substrates with enzymes. This dogma was not always self-evident, but evolved from the remarkable insights of early scientists exploring a number of fundamental living processes.

Although Claude Bernard (1813-1878) never used the term *receptor*, he pioneered a pattern of scientific investigation that permitted clarification of the specificity and selectivity of drug action, particularly in regard to the *locus* of a drug effect. Bernard had a very unpretentious question: he simply wanted to know how the arrow poison curare worked. It was effective when "administered" by an arrow but, interestingly (at least to Bernard), was ineffective when taken by mouth. His early studies explained the importance of the route of administration of this drug for its lethal effects by demonstrating that although curare was unaltered functionally by saliva, gastric juice, bile, or pancreatic juice, it was not absorbed by the gastrointestinal tract, thus accounting for its harmlessness when swallowed. Bernard then wanted to understand just how curare effected its lethal paralysis. It was his impression from general observations that curare did not affect the sensory nerves, but instead altered motor nerve function. By an ingenious group of experiments, he determined that curare blocked the ability of motor nerves to control muscular contraction. Bernard noticed that, after injecting curare under the skin on the back of the frog, the frog showed progressively fewer reflex movements. If he skinned the hind legs of the frog that had been exposed to curare and isolated the lumbar nerve, he could produce no contraction of the leg muscles by stimulating the nerve electrically, whereas he could produce violent contractions if the same electrical stimulus were applied directly to the muscle. Bernard concluded from these experiments that muscle contractility is distinct from the nervous system that produces it and that curare removes the neural control of muscular function (cf. Bernard [1856]).

Bernard did not talk about receptors *per se*, but he did demonstrate that the ability of a drug to elicit its effects depends on its access to a particular location. As a result of his findings, Bernard encouraged investigators not to focus studies of drugs on organs but on organ *systems*, for example, the nervous system or the muscular system. Similarly, he believed that the mechanism of drug toxicity would be better elucidated by focusing on the drug-mediated death of these organ systems, rather than on the death of the organ itself. His own experiments revealed the existence of a neuromuscular "junction" prior to the demonstration of the muscular endplate as a discrete anatomic structure.

It may have been a physicist, rather than a physician or biological scientist, who first provided evidence for *molecular interactions* between two substances that had physiological consequences; Stokes (1864) observed that spectral changes occurred when oxygen was removed from, or subsequently reintroduced to, blood, implicating a complex between oxygen and hemoglobin. However, the biological concept of receptors is generally attributed to Paul Erhlich (1854-1915), although the word **receptor** (receptive substance) was coined by one of Erhlich's contemporaries, J. N. Langley. Erhlich was a remarkable individual whose scientific career spanned (and even spawned) several biomedical disciplines. One overriding principle was common to all of Erhlich's investigative endeavors, and that was **selectivity.**

Erhlich's earliest work involved the distribution of lead in the body, particularly its preferential accumulation in the central nervous system. He had been inspired by a publication of Heubel on lead poisoning, which demonstrated that there were significant differences in the amount of lead found in various organs of animals that had succumbed to lead poisoning. When Heubel exposed the isolated organs of normal animals to dilute solutions of lead, the organs demonstrated the same differential uptake of lead as had been noted *in vivo*. In Erhlich's continuation of these studies, he realized that it was impossible to use a microscope to determine the basis for this differential selectivity of lead uptake in different tissues. Consequently, he changed his experiments to investigate the differential staining of tissues by dyes, as this could be easily detected. He continued to pursue the question of the basis for selectivity, from a more general standpoint. Erhlich's studies on dye distribution originated the concept of "vital staining," and his morphological distinction of leukocytes as acidophilic, basophilic, neutrophilic, or non-granular (based on the relative uptake of dyes of varying chemical constitution) is still in practice today. It was Erhlich's impression that although staining of dead tissue gave information regarding its anatomical structure, the staining of live tissue (i.e., "vital staining") provided insight into the properties and functions of living cells.

Erhlich's most acclaimed studies were his subsequent experiments in immunochemistry, cited as the basis for the Nobel Prize in Medicine awarded to him in 1908. By neutralizing the activity of toxins following incubation of toxins with anti-toxins in a test tube, Erhlich demonstrated that antigen-antibody interactions are direct chemical encounters and not generalized phenomena requiring the biological processes ongoing in a whole animal. From these observations Erhlich developed his "side chain theory" to explain the chemical basis for the immune response. He described the antigen as possessing two active areas: the haptophore (which functioned as the anchorer) and the toxophile (which functioned as the poisoner). He postulated that mammalian cells possess "side chains" that are complementary to certain chemical groups on the haptophore domain of the antigen, and thus serve as the basis for "anchoring" the antigen to the cell. This side chain-haptophore interaction thus gives the "toxophile" portion of the antigen access to cells that possess the appropriate side chains. Pictures reproduced from Erhlich's original notebooks show the side chains drawn with -NH_2 and -SH moieties, thus underscoring his assertion that the basis for these selective interactions between antigen and antibody was a chemical one. Quite clearly, his side chain theory also could explain earlier observations concerning the preferential uptake of lead into the central nervous system and the principle governing vital staining of living cells. Erhlich conjectured that the normal function of cellular side chains was the binding of cell nutrients, and that the affinity of toxic substances for these groups was the fortuitous analogy between the structure of the exogenous toxic substance and the endogenous nutrient. Inherent in Erhlich's side chain theory was the burgeoning concept of specific cell surface receptors as the basis for targeting bioactive agents to the appropriate cell for response.

Erhlich turned his attention from large molecules, such as toxins, to low molecular weight molecules in a series of investigations that earned him recognition as the "father of chemotherapy" (see Albert [1979]). He believed that since the pharmaceutical industry could produce a number of small molecules (e.g., analgesics, antipyretics, and anesthetics) which appeared, at least functionally, to differentiate among various tissues in human beings, it also should be possible to design small molecules that differentiated between human beings and parasites (Erhlich [1913]). His initial studies pursuing this postulate shifted from the protozoan (*Trypanosoma*) to the bacterium (*Treponema*) when Hata showed that the latter organism could produce syphilis in rabbits. Thus, with a model system allowing more detailed studies of chemotherapeutic principles, Erhlich invited Hata to leave Tokyo and join him as a colleague in Frankfurt. Erhlich realized that a particular organism (i.e., *Trypanosoma* versus *Treponema*) was not critical for furthering his studies, because the basis of his experiments on differentiating host from parasite relied only on a *general* principle: that the parasite, as an incessantly motile organism, had a higher rate of metabolism than its host and presumably would be differentially sensitive to the toxic effects of arsenicals. Erhlich's work with a family of arsenical compounds revealed that agents were never entirely specific for the parasite (i.e., he never found his "magic bullet") and, at increasing concentrations, all agents studied had deleterious effects on the host. As a result of this finding, he introduced the term **chemotherapeutic index**, which he defined as the ratio of the minimal curative dose to the maximal tolerated dose. Second, Erhlich maintained that the haptophoric and toxophilic principles that guided immunochemistry also pertained in chemotherapy. Thus, he believed that small molecules also possessed distinct domains for binding to the target cell versus taking part in cellular nutrition or respiration. His own studies established that the arsenoxide group of arsenicals was essential for the lethal effect of these agents and that the chemical substituents on the arsenoxide group were responsible for uptake of the agent. The need first to "bind" the arsenical explained the basis for resistance to arsenicals by particular strains of trypanosomes, i.e., these strains were unable to recognize certain substituents on the phenyl ring attached to the arsenic.

All of Erhlich's studies on the basis of selectivity often are distilled into his often-quoted dictum, *corpora non agunt nisi fixata* (agents cannot act unless they are bound). Consequently, Erhlich's own advice regarding the pursuit of chemotherapeutic agents was to focus on the haptophore group, as it was the *conditio sine qua non* for therapeutic action.

J. N. Langley (1852-1926), of Cambridge University, was a contemporary of Erhlich who studied the chemical basis for autonomic transmission and neuromuscular communication. Langley extended Bernard's studies, which identified curare as a blocker of neuromuscular transmission, by demonstrating that curare also blocked chemical stimulation of the frog gastrocnemius muscle by nicotine, even after severance and degeneration of its motor nerves. However, even under curare "blockade" direct electrical stimulation of denervated muscle could elicit contraction. The mutually antagonistic effects of curare and nicotine, as well as the ability of direct electrical stimulation of the muscle to bypass the effects of curare, led Langley to conclude that nicotine and curare act on the same substance, which is neither nerve nor muscle. Langley called this postulated substance the "receptive substance" (Langley [1909]). The concept of mutual antagonism implying a common site of action was noted by Langley as well as by other contemporaries (e.g., Luchsinger in 1877 and after) for the effect of pilocarpine (agonist) and atropine (antagonist) on contraction of the heart (1909) and on secretion of saliva from the submaxillary gland of the dog (1878). Luchsinger was the first to apply the term "mutual antagonism" to the observed counter-regulatory effects. (See Langley [1878] for a translation from the German of Luchsinger's results and interpretations.) However, Langley emphasized that mutual antagonism depended on the relative concentrations of drugs added and that it had limits. For example, he observed that if he applied extremely large doses of pilocarpine to the artery of the submaxillary gland, secretion was blocked, i.e., pilocarpine could be made to mimic the physiological effect of atropine. Langley also realized that limits to mutual antagonism might be dictated not only by the properties of the receptive substance but also by other secondary effects of the drugs, such as drug-elicited changes in blood flow.

In summarizing his experimental findings, Langley concluded that the effects of the drugs he had observed could reasonably be assumed to result from the existence of some substance(s) in the nerve endings or glands with which both atropine and pilocarpine are capable of forming "compounds." He further postulated that these compounds (complexes) are formed according to some law by which the relative concentration of the drugs and their affinity for the receptive substance are critical factors. Thus, Langley first stated the concept of drug-receptor interaction and predated the algebraic description of these interactions as a consequence of mass action law. Langley observed that the height of the contraction elicited as a result of nicotine interacting with a receptive substance depends on the *rate* of combination of nicotine with this substance as well as the duration of the resulting contraction, and that "saturable" effects on contractility could be observed. Langley actually postulated that if the combination of nicotine with the receptive substance were slow enough and the duration of contraction brief enough, a complete saturation of the receptive substance might occur without eliciting a visible contraction (Langley [1909]).

Despite persuasive evidence that receptors that are specific for particular drugs or endogenous substances do exist and thus determine the selectivity of biological responses to these agents, not all contemporaries or successors of Erhlich and Langley concurred. H. H. Dale (1875-1968) believed that the differential effectiveness of adrenaline analogs in mimicking sympathetic functions in varying tissues could be due to a chemical process, and did not necessarily imply the existence of specific chemical receptors on target tissues. He stated in 1910 that it was equally probable that the limiting factor determining the selective response to various substances might be the ease with which those substances reached their site of action. Thus, he appeared to favor the distributive rather than the interactive properties of a drug in determining its target cell selectivity, although Dale himself acknowledged that his own results could provide no decisive evidence one way or the other (cf. Dale [1914]).

OCCUPANCY THEORY

A. J. Clark (1885-1941) introduced a more quantitative approach to the description of receptor selectivity and saturability (Clark [1926a,b]). Based on his studies of antagonism between acetylcholine and atropine in a variety of muscle preparations, Clark postulated that drugs combine with their receptors at a rate dependent on the concentration of drug and receptor, and that the resulting drug-receptor complex breaks down at a rate proportional to the number of complexes formed (Clark [1927]). This statement implied that drug-receptor interactions obey the principles of mass action and thus could be described by the same isotherms used by Langmuir to describe adsorption of gases onto metal surfaces. Based on Clark's principles, a mathematical expression can be provided to describe drug-receptor interactions:

$$\text{rate of combination} = k_1 A\,(1 - Y) \tag{1.1}$$

$$\text{rate of dissociation} = k_2\,Y \tag{1.2}$$

where k_1 = rate *constant* for combination
k_2 = rate *constant* for dissociation

A = concentration of agonist drug
Y = proportion of receptors occupied by the agonist drug

As will be described in further detail in chapter 2, J. H. Gaddum later extended this mathematical relationship to describe and analyze the competitive antagonism between adrenaline and ergotamine in the rabbit uterus (Gaddum [1926, 1937, 1957]).

At equilibrium, the rate of combination equals the rate of dissociation:

$$k_1 A\,(1 - Y) = k_2 Y$$

and

$$\frac{k_1}{k_2} = \frac{Y}{A(1-Y)}$$

defining K, the equilibrium association constant, as k_1/k_2, and rearranging the above relationship yields

$$Y = \frac{KA}{1+KA} \quad (1.3)$$

Equation 1.3 relates the concentration of drug applied, A, to the proportion of receptors occupied by the drug at equilibrium, Y. This algebraic relationship describing fractional receptor occupancy as a function of drug concentration is analogous to the quantitative relationships between enzyme and substrate introduced by Michaelis and Menten.

RELATIONSHIP BETWEEN OCCUPANCY AND RESPONSE

A. J. Clark extended his hypothesis about the relationship between occupancy and response by postulating that the fraction of receptors occupied, Y, was directly proportional to the response of the tissue. To substantiate this postulate, Clark provided evidence from his studies on acetylcholine-induced contraction of isolated frog *rectus abdominis* muscle and acetylcholine-inhibited contraction of electrically stimulated frog ventricular muscle. If receptor occupancy correlated linearly with receptor-mediated response, then the above equations made certain predictions of what would be expected for the slope of log concentration-response relationships.

Since $K = \frac{Y}{A(1-Y)}$, then rearrangement yields

$$A = \frac{Y}{(1-Y) \bullet K}$$

and predictions could be made about the ratio of drug concentrations eliciting *x*% versus *y*% of response. For example, Clark often compared the ratio of [drug] eliciting 16% versus 84% of a maximal response. If the fraction of receptors occupied correlates directly with the maximal response elicited, then the ratio of drug concentration eliciting 16% versus 84% of maximal response should be around 28 fold, as shown algebraically below:

$$\frac{A_{.84}}{A_{.16}} = \frac{\frac{.84}{.16K}}{\frac{.16}{.84K}} = 28$$

Although some early data of Clark and others describing concentration-response relationships in various contractile systems were consistent with the postulate that the fraction of receptors occupied (implied to be equivalent to the dose of drug added) correlated directly with the fractional response elicited, certain data conflicted with this straightforward relationship between occupancy and effect. First, the slope of the concentration-response relationships reported often was steeper (although sometimes shallower) than predicted from equation 1.3. Second, a number of examples existed in which application of supramaximal concentrations of stimulatory agents did not elicit a maximal contractile response. The latter findings suggest that even saturating occupancy of a receptor population by certain agonist agents might not necessarily elicit a maximal physiological effect (Clark [1937]).

Comparing dose-response relationships for a homologous series of drug analogs often revealed that some agents in the series failed to elicit the same maximal effect, even at supramaximal concentrations. Raventos and Clark (1937) and later Ariëns (1954), Stephenson (1956) and others observed that a *dualism* of behavior was noted for compounds in a homologous series of quaternary ammonium salts in a variety of muscle preparations. These salts had the basic structure:

$$(CH_3)_3\overset{+}{N}-R$$

When the substituent, *R*, was butyl or corresponded to lower members of the series, a maximal muscle contraction was elicited. In contrast, only a weak contraction could be elicited by hexyl and heptyl analogs. Furthermore, the hexyl and heptyl analogs behaved as antagonists when applied to the muscle simultaneously with the butyltrimethylammonium compound.

Ariëns found a similar dualistic behavior of phenylethylamines (chemically related to epinephrine) in elevating blood pressure in decapitated cats. Ariëns drew attention to this enigma: How can a substance which is postulated to interact with a single receptor nonetheless elicit both agonistic and antagonistic effects? He introduced the term **intrinsic activity** to describe the ability of an agent to elicit its pharmacological effect. He expressed the relationship between the agonist effect (E_A) elicited by drug D and the concentration of drug-receptor complexes (DR) as:

$$E_A = \alpha[DR] \tag{1.4}$$

and defined α as the "proportionality constant" or "intrinsic activity" of the particular drug, where intrinsic activity was meant to be a constant determining the effect elicited per unit of DR complex formed. Ariëns still did not alter the fundamental principles of A. J. Clark in his *initial* definition of intrinsic activity. The maximal effect of a given drug still required occupancy of the entire receptor population. The only nuance was that some drugs, even at maximal occupancy, might elicit a biological effect less than that considered to be "maximal" for the system under study. Consequently, this early definition of intrinsic activity proposed by Ariëns addressed the anomalous observation that apparently maximal receptor occupancy by some agonists did not elicit a maximal response. However, this conceptualization still could not explain dose-response relationships that were steeper than predicted by mass action law.

R. P. Stephenson (1956) introduced a major conceptual advance in understanding the quantitative relationship between receptor occupancy and receptor-elicited effects. Stephenson argued that even A. J. Clark's own experimental findings were not in accord with a linear relationship between occupancy and effect. Stephenson concurred that equation 1.3 ($Y = \frac{KA}{1+KA}$) is the probable relationship between the concentration of drug introduced and the concentration of drug-receptor complexes formed. However, Stephenson argued that there was no experimental justification for extending this relationship by supposing that equation 1.3 describes a general relationship between the concentration of drug added and the *response* of the tissue. R. F. Furchgott (1955, 1964) also emphasized that a non-proportionality between occupancy and response was commonly observed. When Stephenson tabulated the slopes of concentration-response curves already reported in the literature, he observed that these slopes typically were steeper than would be predicted if the percentage of maximal response elicited were to correspond directly to the percentage of receptors occupied.

Stephenson (1956) postulated three principles governing receptor-mediated functions that could explain the previous anomalous observation that agonist-response curves often were steeper than the dose-response relationships predicted by simple mass action law. In addition, Stephenson offered an explanation for the observed progressive variation in the agonistic properties of a homologous series of drugs.

1. A maximum effect can be produced by an agonist when occupying only a small proportion of the receptors.
2. The response is not linearly proportional to the number of receptors occupied.
3. Different drugs may have varying capacities to initiate a response and consequently occupy different proportions of the receptors when producing equal responses. This property is referred to as the **efficacy** of the drug. In this setting, a pure competitive antagonist would have zero efficacy.

Stephenson described the relationship between occupancy and response as follows:

S = stimulus given to the tissue
$S = e \cdot y$

where e= efficacy
y = fractional receptor occupancy
R = response of a tissue and $R = f(S)$

indicating that the response is some function (albeit quantitatively unknown) of stimulus S.

If $S = e \cdot y$, then, by mass action law (cf. equation 1.3),

$$S = \frac{eKA}{1+KA} \tag{1.5}$$

Stephenson stated that for an "active agonist," i.e., one with high efficacy and having to occupy only a small portion of the receptors to elicit a maximal response, KA would be small relative to 1. In this situation, equation 1.5 reduces to:

$$S = eKA \tag{1.6}$$

This definition of efficacy is distinct from that originally proposed by Ariëns. However, Ariëns later changed his definition of intrinsic activity to one formally equivalent to this efficacy term of Stephenson (Van Rossum and Ariëns [1962]).

To test the validity of his postulates regarding various efficacies for different agonists, Stephenson carried out two separate lines of investigation. First, he evaluated the concentration-response for the "full agonists" (which he called "active agonists") acetylcholine and histamine in eliciting contraction of the guinea pig ileum. Stephenson quantitated these data based on the ratios of drug concentrations needed to elicit certain graded responses. Based on Clark's hypothesis, for example, the ratios of the concentration of agonist eliciting 80% versus 20% contraction should be 16 (see earlier algebraic determination of these concentration ratios), and those for 20% versus 50% and 50% versus 80% contraction should be 4. Stephenson noted that the values he obtained were considerably less than the predicted values of 4, and noted this same discrepancy when he calculated agonist ratios from contractile data already published in the literature. (An exception was the concentration-response relationship of adrenaline for contracting rabbit aorta strips published by Furchgott and Bhadrakom in 1953.) Stephenson thus concluded that many agonists elicit a far greater contractile response than would be predicted based on the extent of receptor occupancy.

In a second series of experiments, Stephenson studied the series of alkyl-trimethylammonium salts, introduced by Raventos and Clark, on contraction of the guinea pig ileum. He noted that the lower homologs (e.g., butyl-trimethylammonium) behaved like acetylcholine, an agonist, whereas higher homologs acted like atropine, an antagonist. He interpreted this antagonism as a property expected for a drug with low efficacy. Thus, the drug produces a response much lower than maximal even when occupying all or nearly all of the receptors. However, because a drug with low efficacy can nonetheless occupy the receptors, it decreases the response elicited by a drug with high efficacy when added simultaneously. Stephenson termed these low-efficacy drugs "partial agonists" because they possessed properties intermediate between agonists and antagonists. (These partial agonists are what Ariëns [1954] referred to as drugs with a dualism of action or mixed agonists/antagonists.) The ability of partial agonists to antagonize agonist effects formed a basis for determining the affinity of partial agonists for the receptor. This methodology will be described in further detail in chapter 2.

CONCEPT OF SPARE RECEPTORS

The finding that some agonists could elicit maximal physiological effects by occupying only a small fraction of the total receptor population suggested that there were "spare receptors." Avraim Goldstein (1974) offered a tenable teleological explanation for such a phenomenon. In circumstances where the desired response must be rapid in onset and in termination (as in neurotransmission), a spare receptor capacity provides a mechanism for obtaining a response at a very low concentration of an agonist that nonetheless has a relatively low affinity for the receptor. Sensitivity to low drug concentrations is achieved by the spare receptor capacity. The low affinity (i.e., low K_A) of the drug assures its more rapid rate of dissociation, since $K_A = k_1/k_2$. Alternatively, if sensitivity to low concentrations of agonist were achieved by a high affinity of the drug for the receptor, then the rate of reversal of the effect would necessarily be slow.

Documentation of the existence of spare receptors, however, came not from studies of agonist concentration-response profiles but instead from studies of receptor antagonism. Several examples of so-called anomalous antagonism had been described that simply could not be explained by A. J. Clark's hypotheses or by the equations describing simple competitive antagonism introduced by Gaddum. To evaluate the nature of a particular drug's antagonistic effects, agonist concentration-response curves were obtained in the presence of increasing concentrations of the antagonist. A rightward parallel shift of these curves was consistent with reversible competitive antagonism, and estimates of receptor affinity for the antagonist could be obtained by the method of Schild (see chapter 2) or by Lineweaver-Burk plots, as had been popularized in enzyme kinetic studies. However, as pointed out by M. Nickerson (and other contemporaries who obtained similar findings in other systems), one occasionally could obtain evidence consistent with reversible competitive antagonism when other data nonetheless suggested that reversible competitive interactions were not a likely explanation for the nature of the antagonism (see Furchgott [1955]). For example, β-haloalkylamines, such as dibenamine, were known to block histamine and catecholamine receptors irreversibly, since blockade of contraction by β-haloalkylamines never could be reversed despite extensive washing of the isolated tissue preparation. Except at higher concentrations of these antagonists, however, data for the blockade of histamine-induced contractions resembled that expected for reversible, competitive antagonism: a shift to the right of the agonist concentration-response curve with no change in the slope of the curve or the maximal effect elicited. Only at high concentrations of β-haloalkylamines was a decrease in both the slope and maximal effect of the agonist finally detected for histamine-induced effects. Nickerson is credited with explaining these anomalous antagonisms by demonstrating that receptor occupancy is not necessarily the limiting factor in tissue activation, i.e., that spare receptors exist. As an example, Nickerson demonstrated in 1956 that occupancy of only 1% of the histamine receptor population in guinea pig ileum was required to elicit maximal contractile effects, suggesting the existence of a large receptor reserve for histamine receptors in this tissue. Receptor reserves were not always so dramatic, however. Furchgott (1955) noted that for epinephrine there was only a shift of half a log unit, if anything, before a decrease in maximal response was observed following β-haloalkylamine exposure.

The impact of receptor reserve, or "spare receptors," on the sensitivity of a system to agonist is most readily (and dramatically) revealed in heterologous receptor systems where receptor density can be controlled in a straightforward fashion. Here, increases in receptor expression often are noted to be paralleled by a decrease in the concentration of agonist eliciting 50% of maximal response, defined as EC_{50} (e.g. Whaley et al. [1994]). As predicted by receptor theory, the efficacy of partial agonists *also* is increased as receptor density is increased (Tan et al. [2003]).

OPERATIONAL MODELS OF PHARMACOLOGICAL AGONISM

Black and Leff (1983) developed a mathematical model, dubbed the **operational model for agonism**, in an effort to provide quantitative descriptors for the frequently observed nonlinear relationship between occupancy and response, or effect (cf. figure 1-2). This model assumes that agonist A binds to receptors R in a bimolecular reaction obeying mass action law, such that:

$$[AR] = \frac{[R_o][A]}{K_A + [A]} \tag{1.7}$$

where R_o=total receptor concentration

K_A=equilibrium association constant M^{-1}, the reciprocal of which defines affinity.

The relationship given in equation 1.7 takes the form of a rectangular hyperbola: $y = mx/(a + b)$. Thus, a plot of $[A]$ on the x axis versus $[AR]$ on the y axis will resemble a rectangular hyperbola.

To find a numerical function that relates the concentration of agonist · receptor complex $[AR]$ to observed *effect* E, Black and Leff posited that E itself is a rectangular hyperbolic function of $[AR]$, such that:

$$\frac{E}{E_m} = \frac{[AR]}{K_E + [AR]} \tag{1.8}$$

Where E_m= maximal effect or response
E = effect elicited at a given level of occupancy, i.e., $[AR]$
K_E= value of $[AR]$ that elicits the half-maximal effect

If equations 1.7 and 1.8 are combined, then

$$\frac{E}{E_m} = \frac{[R_o][A]}{K_A K_E + ([R_o]+[K_E])[A]} \tag{1.9}$$

Assuming that receptor occupancy can be described by a rectangular hyperbolic expression, the analysis of Black and Leff demonstrates that the transducer function, i.e., that function(s) linking occupancy to response, must be hyperbolic if the observed $E/[A]$ relationship is hyperbolic. An important component defined in this model is dubbed the **transducer ratio, τ** (tau),

$$\tau = \frac{[R_o]}{K_E}$$

which measures the efficiency by which occupancy of the receptors is transduced to a biological effect. It can be seen that τ reflects properties of the tissue, namely concentration of receptors, and the consequences of drug-receptor interaction, namely, the potency of an agonist in eliciting a response due both to receptor affinity for the agonist and receptor efficiency in translating receptor occupancy to response. Consequently, the operational agonism model introduced by Black and Leff resembles the conceptual framework of Furchgott, acknowledging both tissue and drug receptor properties as contributors to the ultimate response (equation 2.2 and related text). However, instead of the empirical nature of the efficacy constant, the value of τ can be measured as an experimentally observed relationship between occupancy and response.

The premise under which Black and Leff undertook their mathematical description of $E/[A]$ relationships was to provide quantitative descriptors of experimentally observed $E/[A]$ curves. To provide descriptors for experimental $E/[A]$ curves that were steeper or shallower than predicted for a rectangular hyperbola, equation 1.8 can be converted to a logistic form:

$$E = \frac{E_m[AR]^n}{K_E + [AR]^n} \tag{1.10}$$

where the rectangular hyperbola represents the special case wherein the experimental term n equals 1; steeper curves have values of $n > 1$ and shallower curves have values of $n < 1$.

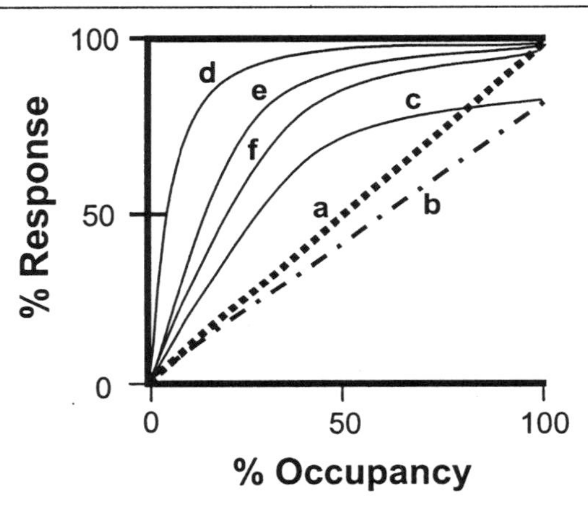

***Figure 1-2*. Relationships of occupancy to response inherent in the evolution of receptor theory.** A. J. Clark first proposed a direct proportional relationship between occupancy and response (a). Clark Furchgott and Ariëns, to name a few, found that this direct proportional relationship was not always reflected by the data obtained; Ariëns first proposed the concept of intrinsic activity, and his concepts are described by line b, which has a slope > 0 < 1.0. Stephenson and others realized a more complex relationship of occupancy and response, so that partial agonists could either be agents that did not elicit a maximal effect even at full occupancy (c) or agents that ultimately elicited a full, maximal response did so by activating a larger fraction of the receptor population (e & f) than a native agonist, or "full", "active" agonist drug (d).

This model of operational agonism has been useful in predicting the behavior of rectangular hyperbolic and non-hyperbolic $E/[A]$ curves when changes in R_o occur, such as when an investigator examines the relative order of agonist potency in tissues with differing receptor reserve (Black, Leff and Shankley, with an appendix by Wood [1985]) or with receptors distributed into differing receptor states, due to allostery or ternary complex formation

(Black and Shankley [1990]). In addition, although this model (and its inherent algebraic descriptions) offers no insight into the molecular events linking [*AR*] to *E* in a given system, the term τ *does* provide a quantitative descriptor of an effector response to increasing concentrations of an agonist that is useful in comparing the properties of that system, for example, before and after desensitization (Lohse et al. [1990]), such that the entire *E*/[*A*] curve is taken into account, rather than simply comparing response at a single concentration of agonist.

RATE THEORY

Inherent in all of the postulates described thus far is the assumption that the number of receptors *occupied* somehow determines the response observed, a conceptual framework referred to as **occupancy theory**. W. D. M. Paton (1961) explored a unique hypothesis, **rate theory**, to explain drug action in an attempt to provide a theoretical basis for some experimental findings he and earlier investigators had reported that were inconsistent with any extant theories relating drug effect to receptor occupancy. Although rate theory today does not appear to explain well-characterized receptor-dependent phenomena, the conceptual development of rate theory and the algebraic descriptors of a model for rate theory are briefly given here to emphasize how the assumptions inherent in developing a model for receptor-mediated response have a dramatic consequence on the ultimate equations describing that response. The findings that Paton could not reconcile with occupancy theory were: 1) the observation that excitation by certain agonists (such as nicotine) often was followed by a "block" in receptor function, 2) the trace stimulant actions of certain antagonists, and the persistence of these effects, and 3) that the effects of agonists often demonstrated a "fade" with time. Paton postulated that excitation was proportional to the *rate* of drug-receptor interaction, rather than to the number of receptors occupied by the drug. He visualized excitation as resulting from the *process* of occupation of the receptor, not occupation itself, and that each association event between drug and receptor resulted in one "quantum" of excitations. The algebraic description of rate theory, and its comparison to occupancy theory, are as follows:

definitions:

x(g/ml)	concentration of drug added to bath
p	proportion of receptors occupied at time t, seconds
$A(sec^{-1})$	association rate/receptor, equivalent to $k_1 \cdot x \cdot (1 - p)$
$k_2p(sec^{-1})$	dissociation rate/receptor
$k_1(sec^{-1}\ g^{-1}\ ml)$	association rate *constant*

k_2(sec^{-1})	dissociation rate *constant*
k_e(g/ml)	equilibrium dissociation constant, equivalent to k_2/k_1
y(mm)	response recorded experimentally
	occupation theory = $\phi' p$
	rate theory = ϕA
f (fade ratio)	ratio of equilibrium plateau response to initial peak response

The proportion of receptors occupied at equilibrium (p) can be described by the relationship:

$$p = \frac{k_1 x}{k_2 + k_1 (x)} = \frac{x}{x + k_2 / k_1} \qquad (1.11)$$

This follows from the definition that, at equilibrium, the rate of association equals the rate of dissociation, i.e., $k_1 \cdot x\,(1 - p) = k_2 p$.

If the response at equilibrium is proportional to occupation, then:

$$y = \phi' p$$

where ϕ' is a constant that includes the efficacy factor (e) of Stephenson.[1] If, however, the equilibrium response elicited is proportional to the rate of receptor occupancy, then:

response $y = \phi A$

since

$$A = k_1 \cdot x \cdot (1 - p)$$

At equilibrium, A also = $k_2 p$, and substituting for p as in equation 1.12:

$$A = k_2 \cdot \frac{x}{x + k_2 / k_1} \qquad (1.12)$$

and response $y = \dfrac{\phi k_2 x}{x + k_2 / k_1}$

---

[1] Paton referred to the relationship $y = \phi' p$ as "occupation theory," and conceptualized the constant ϕ' as a constant that includes the factor α of Ariëns or the factor e of Stephenson as well as a factor linking the intensity of chemical stimulation with the recorded response. However, ϕ' should not be equated with the efficacy factor of Stephenson, as Stephenson demonstrated that response is not necessarily proportional to occupation.

To compare equilibrium effects obtained via occupancy versus rate theory:

Occupancy theory:

$$y = \phi' \frac{x}{x + k_2 / k_1}$$

Rate theory:

$$y = \phi \frac{k_2 x}{x + k_2 / k_1}$$

and the difference between the two responses is k_2!

It is interesting to compare the observations one would predict if response were attributable to the *rate* versus the *extent* of receptor occupancy. Paton noted that occupancy theory predicted that, prior to equilibrium, the response observed experimentally should rise to a plateau and do so with a time constant of $k_1x + k_2$. This plateau corresponds to the equilibrium, or steady state, response. For rate theory, in contrast, drug action should be its highest at the outset (since response = ϕA, $A = k_1 \cdot x \cdot (1 - p)$, and p is infinitesimal at early time points) and that the initial response would decrease to an equilibrium plateau $(k_2x/(x + k_2/k_1)$, with a time constant of $k_1x + k_2$. This decline in response from the first peak to a later plateau was referred to as **fade**. The time constant $(k_1x + k_2)$ described two entirely different phenomena for occupancy theory versus rate theory. For occupancy theory, $(k_1x + k_2)$ measures the rate of rise of response, whereas for rate theory $(k_1x + k_2)$ measures the time for decline from maximal response. Consequently, rate theory predicts a fade to occur for all compounds in which k_2 is other than very large compared to k_1x.

ALLOSTERIC THEORY

The preceding discussions of efficacy, receptor reserve, and rate theory represent attempts to describe quantitatively the general observation of nonlinear coupling between receptor occupancy and response (cf. figure 1-2). Another theoretical framework adopted for this same purpose is that of **allostery**. Earlier studies of A. V. Hill and others on the binding of oxygen to hemoglobin indicated that the binding of oxygen to each of the four heme moieties of the hemoglobin tetramer did not obey the principles of mass action. Instead, binding of the first oxygen facilitated binding of the second, etc., such that the resulting oxygen saturation curve was considerably steeper than that predicted by mass action law. Hemoglobin is an example of an allosteric system. Monod, Wyman, and Changeux (1965) proposed a model (referred to as the MWC model) that could account for allosteric phenomena. Since they observed that most allosteric systems were oligomers involving several identical subunits, their model assumes the existence of an oligomeric protein. The following statements paraphrase the allosteric MWC model proposed in 1965, which has narrow and strict definitions of the properties of what they defined as an allosteric system.

1. Allosteric proteins are oligomers. The protomers are associated such as they are all functionally equivalent.
2. There is only one site for binding of each ligand on each protomer.
3. The conformation of each protomer is constrained by its association with the other protomers.
4. There are at least two "states" reversibly accessible to allosteric oligomers, described by the symbols R and T.
5. The affinity of one (or more) binding site toward its specific ligand is altered when a transition occurs from one state to another.

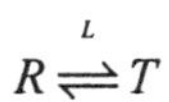

The R and T states are assumed to be in equilibrium in the absence of ligand. An allosteric ligand, denoted as F, is one that possesses a different affinity for the two accessible states and thus displaces the equilibrium of the two states to a new equilibrium favoring the state with higher affinity for F. If the two states correlate with different functional consequences, then F can be seen to influence the expression of function by determining the ratio of oligomers in the R and T states. The equilibrium constant for the R $\rightleftharpoons$ T transition is denoted as L, and is referred to as the **allosteric constant.** When this value is very large (due to an extreme difference in the affinity of F for the R or T state), then the "cooperativity" noted is extremely marked, and the entire enzyme or receptor system may behave as if it has only one "state," i.e., active or inactive.

Nicotinic cholinergic receptor-mediated Na^+ influx and membrane depolarization represent biological responses consistent with this model. The steep dose-response effects of nicotinic cholinergic agents on membrane depolarization could be accounted for in terms of allosteric theory (see Karlin [1967]; Changeux and Podleski [1968]; Colquhoun [1973]). The cholinergic receptor was postulated to exist in two interconvertible states, a depolarized or "active" state (D) and a polarized or "inactive" state (P). Agonists were postulated to have a higher affinity for the D state, whereas antagonists were postulated to have a much higher affinity for the P state. In this model, antagonists would be predicted to shift the equilibrium toward an inactive state by preventing the shift to the D state. Partial agonists would be predicted to have variable affinities for the two states but a preferential affinity for the D state. Differences in maximal response elicited by partial agonists would then be interpreted to result from a portion of the receptor population remaining in the P state, even in the face of maximal receptor occupancy.

To generalize this model, however, it is useful to categorize the MWC allosteric model as a **two-state model**. Generally, the term *allosteric* describes binding to and altering receptor regulated responses at a site distinct from that which binds endogenous agonist agents, defined as the *orthosteric* site. Allosteric regulation can occur by interactions on the receptor itself, or interactions with another protein (such as heterotrimeric G proteins for G protein-coupled receptors (GPCRs)) that influence receptor conformation(s) and result in propagated changes in binding at the orthosteric site. This concept will be discussed further in chapters 2, 3 and 4, since allosteric sites on receptors regulated by small molecules have considerable importance in drug discovery. Allosteric effects are saturable (Birdsall et al. [1996]), and thus there is a "ceiling" to the effect of these agents when administered therapeutically. Allosteric enhancers could be developed that exert effects only when the *orthosteric*, endogenous agonist is present, further enhancing the selectivity of the drug response. Whereas orthosteric binding sites are highly conserved-making subtype-specific agonists challenging to develop-the surfaces in receptors where allosteric modulators interact may not be so well conserved, and this divergence in sequence homology also may permit highly subtype-specific allosteric modifiers to be developed (Christopoulos and Kenakin [2002]).

BEYOND TWO-STATE RECEPTOR THEORY

A number of experimental observations in native cells (or tissues) as well as in systems expressing heterologous receptors cannot be accounted for by a simple two-state model of an active (R^*) versus inactive (R) receptor, as first proposed by Monod, Wyman and Changeux (above), using the terms R and T for R^* and R, respectively. In the two state model for G protein-coupled receptors, the active R^* form of the receptor would be the form interacting with G proteins. In the absence of agonist, the distribution of receptor between the R^* and R states would be governed by the equilibrium constant L, and the activity of an agonist (A) would be determined by the values of K_A versus $K_A{}^*$, the equilibrium association constant for agonist at R versus R^*, respectively. A restatement of the generalized two-state model follows:

$$\begin{array}{ccccl} A+R^* & \overset{KA^*}{\rightleftarrows} & AR^* & & \text{active} \\ \Updownarrow & & \Updownarrow & & \\ A+R & \overset{KA}{\rightleftarrows} & AR & & \text{resting} \end{array}$$

(1.13)

The two state model (1.13) can account for multiple receptor-G protein coupling events, e.g. if R^* is able to activate more than one G protein. However, in the two-state model, an agonist can only alter the ratio of R to R^* in a proportional way, so that the interactions of R^* with two G proteins would similarly be altered in a proportional way. A two-state model, therefore, cannot account for the observations that different orders of agonist potency (and efficacy) have been reported for a single receptor interacting with different G proteins (Spengler et al. [1993]; Kenakin [1995]; Perez et al. [1996]; Berg et al. [1998]). A **three-state model** (Leff et al. [1997]; Hall [2000]) is the simplest case that can account for these findings.

$$
\begin{array}{cccl}
A+R^* & \overset{KA^*}{\rightleftarrows} & AR^* & \text{active} \\
\Updownarrow & & \Updownarrow & \\
A+R & \overset{KA}{\rightleftarrows} & AR & \text{resting} \\
\Updownarrow & & \Updownarrow & \\
A+R^{**} & \overset{KA^{**}}{\rightleftarrows} & AR^{**} & \text{active}
\end{array}
\qquad (1.14)
$$

In the three state model (equation 1.14), the receptor is distributed among three unoccupied (R, R^* and R^{**}) and three occupied states, where R^* and R^{**} would be predicted to interact with different downstream entities-which, in the case of G protein-coupled receptors, might represent different G proteins. This model is algebraically indistinguishable from the model for sequential binding and conformational stabilization proposed by Gether and Kobilka [1998], except that in the latter formulation the third conformation, defined as R_o, is stabilized by inverse agonists.

In the three state model given in 1.14, agonist activity depends on three equilibrium association constants (K_A, K_A^* and K_A^{**}). Since, by definition, the K_A^* and K_A^{**} constants can vary for each agonist, this model can explain varying efficacies and orders of agonist potencies for two signaling outputs from a single receptor. There are other, less intuitively obvious, consequences of this model, as well. Since there is a finite number of receptors, enrichment of one receptor state (e.g., R^{**}) can occur at the expense of the others, meaning that an agonist that posesses extremely high efficacy through one pathway will necessarily express low efficacy through the other.

Under some circumstances, it is possible experimentally to "isolate" one receptor-activated G protein-coupled pathway from another, e.g. by pertussis toxin treatment to uncouple receptors from G_i or G_o-coupled signaling (Albert and Robillard [2002]) or by G protein-selective disrupting peptides (Gilchrist et al. [2002]). Under these circumstances, the agonist-response curves in the "isolated" conditions correspond to the individual two-state systems that make up the three state model. The potency ratios and efficacy profiles of each agonist will now differ from the unperturbed three-state system, since by disrupting receptor interaction with one versus another G protein, enrichment of R^* and R^{**} states by the different agonists does not mutually deplete other active states.

Investigators are becoming more aware that agonist-independent, or constitutive, activity occurs for many receptor systems in native cells (Seifert and Wenzel-Seifert [2002]), and that this activity can be exaggerated for experimental purposes by overexpression of a given receptor in a heterologous system. In a highly consitutively active system, significant levels of R^* and R^{**} are present in the absence of agonist. Since, as noted above, agonists which enrich R^* do so at the expense of R^{**}, a ligand acting with high agonist efficacy through one pathway will act as an inverse agonist through the other, particularly in systems with sufficient constitutive activity to readily measure suppression of that activity (i.e., inverse agonism). By extension, agonists with only subtle preference for R^* versus R^{**} will enrich both active states without extensive mutual depletion but also act as partial agonists through both the R^* and R^{**}-coupled pathways. These outcomes are modeled in Leff et al. [1997].

Further expansions of this and other models developed to describe and predict observations in G protein-coupled receptor systems are discussed in chapter 4, with particular emphasis on analyzing interactions of agonists versus antagonists with G protein-coupled receptors, as monitored in radioligand binding experiments. However, it is important to note that these more complex models, as well as the biological data, provide a significant modification of the original understanding of partial agonists. Original perceptions, as described by two-state models, anticipated a single agonist-induced or active receptor conformation which effectively transduced signal to its cognate G protein or effector system. Partial agonists were interpreted to evoke this conformation less frequently (i.e., bind to and/or stabilize R^* with lower affinity). Yet, it is clear that multiple active receptor conformations exist, and are uniquely evoked by different agonists. This is evident not only in cellular systems where converse agonist profiles are observed for a single receptor's activation of multiple signaling pathways, but also in the analysis of purified receptor systems. Thus, fluorescent spectroscopic studies of ligand-induced conformational changes in the G protein-coupling domain of the β_2-adrenergic receptor revealed that conformations induced by a full agonist can be distinguished from those induced by partial agonists (Seifert et al. [2001]; Gahnouni et al. [2001]). These investigators propose a model not only with multiple, agonist-specific states but also where activation of downstream events occurs through a *sequence* of conformational changes (see also Gether and Kobilka [1998]; Kobilka [2004]; Liapakis [2004]), adding further complexity to our molecular understanding of physiological and pharmacological phenomena. This complexity, however, is also the basis for developing therapeutic agents that can achieve not only receptor-specific, but also pathway-specific, activation, and hence enhance clinical selectivity.

SUMMARY

This chapter provides a synopsis of the development of the receptor concept to explain differential tissue distribution and ultimate specificity of drug action. A relationship between available drug concentration and the proportion of receptors occupied was quickly advanced which implied that drug-receptor interactions obey mass action principles, in a manner analogous to enzyme-substrate interactions. However, Clark's postulate that the extent of receptor occupancy correlated directly with the extent of elicited response did not explain the majority of experimental findings. To explain the frequent observation that a maximal response could be elicited by occupying only a small fraction of the total receptor population, Stephenson postulated that drugs possess varying efficacies, such that a maximal response can be evoked by occupying differing fractions of the receptor population. Black and Leff have offered a quantitative description of occupancy/response relationships that permits calculation of a transducer ratio for comparison of entire dose-response curves in different tissues or following various experimental manipulations, bringing the quantification of practical descriptors to the nonlinear occupancy-response relationships first introduced by Stephenson. Allosteric models were proposed to account for anomalous antagonisms as well as nonhyperbolic dose-response relationships alluded to earlier. However, multi-state, rather than two-state, models are necessary to describe the varying orders of potencies of a single agonist at a single receptor in evoking different signal outputs.

Distinguishing among the many molecular models that explain how receptor occupancy is linked to biological response ultimately requires purification of the receptor and reconstitution with its purified "effector system," be it ion translocation or modulation of enzymatic activities. Rigorous characterization of the receptor-response system in the intact target cell is a crucial prerequisite for ultimately understanding the molecular basis for the physiological response observed *in vivo*, as it is only to the extent that the purified and reconstituted assembly mimics the native receptor-response system that the *in vitro* system can provide unequivocal insights into receptor mechanisms. Chapter 2 summarizes available methods for determining receptor specificity, the affinity of the putative receptor for its specific agonist, partial agonist, and antagonist agents based on measurements of receptor-mediated response.

REFERENCES

Albert, *A*. (1979) Chemotherapy: History and principles. In *Selective Toxicity* (6th ed.), Chapman and Hall (eds.). New York: John Wiley and Sons, pp. 182-199. (This section of chapter 6 summarized P. Erhlich's fundamental contributions to chemotherapy and general "receptor" principles.)

Albert, P.R. and Robillard, L. (2002) G protein specificity: traffic direction required. Cell Signal. 14(5):407-418.

Ariëns, E.J. (1954) Affinity and intrinsic activity in the theory of competitive inhibition. Part I. Problems and theory. Arch. Int. Pharmacodyn. 99:32-49.

Ariëns, E.J. (1960) Receptor reserve and threshold phenomena. I. Theory and experiments with autonomic drugs tested on isolated organs. Arch. Int. Pharmacodyn. 127:459-478.

Ariëns, E.J. and deGroot, W.M. (1954) Affinity and intrinsic-activity in the theory of competitive inhibition. III. Homologous decamethonium-derivatives and succinylcholine-esters. Arch. Int. Pharmacodyn. 99:193-205.

Barger, G. and Dale, H.H. (1910) Chemical structure and sympathomimetic action of amines. J. Physiol. 41:19-59.

Berg., K.A., Maayani, S., Goldfarb, J., Scaramellini, C., Leff, P. and Clarke, W.P. (1998) Effector Pathway-Dependent Relative Effficacy at Serotonin Type 2A and 2C Receptors: Evidence for Agonist-Directed Trafficking of Receptor Stimulus. Mol. Pharm. 54:94-104.

Bernard, C. (1856) Physiological analysis of the properties of the muscular and nervous system by means of curare. Comptes Rendus Acad. de Sci. 43:825-829. Translated and reprinted in *Readings in Pharmacology*, L. Shuster (ed.). Boston: Little, Brown and Company, pp. 73-81.

Birdsall, N.J., Lazareno, S. and Matsui, H. (1996) Allosteric regulation of muscarinic receptors. Prog. Brain Res. 109: 147-151.

Black, J. W. and Leff, P. (1983) Operational models of pharmacological agonism. Proc. Royal Soc. London B. 220:141-162.

Black, J.W., Leff, P. and Shankley, N.P., with an appendix by J. Wood (1985) An operational model of pharmacological agonism: The effect of *E*/[*A*] curve shape on agonist dissociation constant estimation. Br. J. Pharm. 84:561-571.

Black, J.W. and Shankley, N.P. (1990) Interpretation of agonist affinity estimations: The question of distributed receptor states. Proc. Royal Soc. London B. 240:503-518.

Changeux, J.-P. and Podleski, T.R. (1968) On the excitability and cooperativity of the electroplax membrane. Proc. Natl. Acad. Sci. USA 59:944-950.

Christopoulos, A. and Kenakin, T.P. (2002) G Protein-Coupled Receptor Allosterism and Complexing. Pharm. Rev. 54:323-374.

Clark, A.J. (1926a) The reaction between acetyl choline and muscle cells. J. Physiol. 61:530-546.

Clark, A.J. (1926b) The antagonism of acetyl choline by atropine. J. Physiol. 61:547-556.

Clark, A.J. (1927) The reaction between acetyl choline and muscle cells. Part II. J. Physiol. 64:123-143.

Clark, A.J. (1937) *General Pharmacology*. Berlin: Verlag von Julius Springer, pp. 61-98, 176-206 and 215-217.

Clark, A.J. and Raventos, J. (1937) The antagonism of acetylcholine and of quaternary ammonium salts. Quant. J. Exp. Physiol. 26:275-392.

Colquhoun, D. (1973) The relation between classical and cooperative models for drug action. In *Drug Receptors*, H.P. Rang (ed.). Baltimore: University Park, pp. 149-182.

Dale, H.H. (1914) The action of certain esters and ethers of choline, and their relation to muscarine. J. Pharm. Exp. Ther. 6:174-190.

Erhlich, P. (1913) Chemotherapeutics: Scientific principles, methods and results. Lancet 2:445-451.
Furchgott, R.F. (1955) The pharmacology of vascular smooth muscle. Pharm. Rev. 7:183-235.
Furchgott, R.F. (1964) Receptor mechanisms. Ann. Rev. Pharmacology 4:21-50.
Furchgott, R.F. and Bhadrakom, S. (1953) Reactions of strips of rabbit aorta to epinephrine, isoproterenol, sodium nitrite and other drugs. J. Pharm. Exp. Ther. 108:129-143.
Gaddum, J.H. (1926) The action of adrenalin and ergotamine on the uterus of the rabbit. J. Physiol. 61:141-150.
Gaddum, J.H. (1937) The quantitative effects of antagonistic drugs. J. Physiol. 89:7P-9P.
Gaddum, J.H. (1957) Theories of drug antagonism. Pharm. Rev. 9:211-217.
Gether, U. and Kobilka, B.K. (1998) G Protein-coupled Receptors. II. Mechanism of Agonist Activation. J. Biol. Chem. 273(29):17979-17982.
Ghanouni, P., Bryczynski, Z., Steenhuis, J.J., Lee, T.W., Farrens, D.L., Lakowicz, J.R. and Kobilka, B.K. (2001) Functionally Different Agonists Induce Distinct Conformations in the G Protein Coupling Domain of the β_2 Adrenergic Receptor. J. Biol. Chem. 276(27):24433-24436.
Gilchrist, A., Li, A. and Hamm, H.E. (2002) Design and use of C-terminal minigene vectors for studying role of heterotrimeric G proteins. Methods Enzymol. 344:58-69.
Goldstein, A., Aronow, L. and Kalman, S.M. (1974) *Principles of Drug Action: The Basis of Pharmacology* (2nd ed.). New York: John Wiley and Sons, pp. 82-111.
Hall, D.A. (2000) Modeling the Functional Effects of Allosteric Modulators at Pharmacological Receptors: An Extension of the Two-State Model of Receptor Activation. Mol. Pharm. 58:1412-1423.
Karlin, A. (1967) On the application of a “plausible model” of allosteric proteins to the receptor of acetylcholine. J. Theoret. Biol. 16:306-320.
Kenakin, T. (1995) Agonist-receptor efficacy II: agonist trafficking of receptor signals. TiPS 16:232-238.
Kenakin, T. (2004) Principles: Receptor Theory in Pharmacology. Trends Pharm. Sci. 25:186-192.
Kobilka, B. (2004) Agonist binding: a multi-step process. Mol. Pharm. 65:1060-1062.
Langley, J.N. (1878) On the physiology of the salivary secretion. Part II. On the mutual antagonism of atropin and pilocarpin, having especial reference to their relations in the submaxillary gland of the cat. J. Physiol. 1:339-369.
Langley, J.N. (1909) On the contraction of muscle, chiefly in relation to the presence of “receptive” substances. Part IV. The effect of curare and of some other substances on the nicotine response of the sartorius and gastrocnemius muscles of the frog. J. Physiol. 39:235-295.
Leff, P., Scaramellini, C., Law, C. and McKechnie, K. (1997) A three-state receptor model of agonist action. TiPS 18(10):355-362.
Liapakis, G., Chan, W.C., Papdokostaki, M. and Javtch, J.A. (2004) Synergistic contributions of the functional groups of epinephrine to its affinity and efficacy at the β_2 adrenergic receptor. Mol. Pharm. 65:1181-1190.
Lohse, M.J., Benovic, J.L., Caron, M.G. and Lefkowitz, R.J. (1990) Multiple pathways of rapid $\beta 2$-adrenergic receptor desensitization: Delineation with specific inhibitors. J. Biol. Chem. 265(6):3202-3211, especially appendix pp. 3210-3211.
Monod, J., Wyman, J. and Changeux, J.-P. (1965) On the nature of allosteric transitions: A plausible model. J. Mol. Biol. 12:88-118.
Neubig, R.R., Spedding, M., Kenakin, T. and Christopoulos, A. (2003) Update on Terms and Symbols in Quantitative Pharmacology. NC-IUPHAR XXXVIII 55:597-606.
Nickerson, M. (1956) Receptor occupancy and tissue response. Nature 78:697-698.
Paton, W.D.M. (1961) A theory of drug action based on the rate of drug-receptor combination. Proc. Royal Soc. London B. 154:21-69.

Perez, D.M., Hwa, J., Gaivin, R., Mathur, M., Brown, F. and Graham, R.M. (1996) Constitutive activation of a single effector pathway: evidence for multiple activation sites of a G protein-coupled receptor. Mol. Pharmacol. 49:112-122.

Stephenson, R.P. (1956) A modification of receptor theory. Br. J. Pharm. 11:379-393.

Seifert, R. and Wenzel-Seifert, K. (2002) Constitutive activity of G protein-coupled receptors: cause of disease and common property of wild-type receptors. Naunyn Schmiedebergs Arch. Pharmacol. 366(5):381-416.

Seifert, R., Wenzel-Seifert, K., Gether, U. and Kobilka, B.K. (2001) Functional Differences between full and Partial Agonists: Evidence for Ligand-Specific Receptor Conformations. J. Pharm. & Exp. Ther. 297:1218-1226.

Spengler, D., Waeber, C., Pantaloni, C., Holsboer, F., Bockaert, J., Seeberg, P.H. and Journot, L. (1993) Differential signal transduction by five splice variants of the PACAP receptor. Nature 365:170-175.

Stokes, G.G. (1864) On the Reduction and Oxidation of the Colouring Matter of the Blood. Proc. Roy. Soc. London 13:355-364.

Tan, C.M. and Limbird, L.E. (2003) Heterozygous α_{2A}-adrenergic receptor mice unveil unique therapeutic benefits of partial agonists. Proc. Natl. Acad. Sci. USA 99(19):12471-12476.

Thron, C.D. and Waud, D.R. (1968) The rate of action of atropine. J. Pharm. Exp. Ther. 160:91-105.

Van Rossum, J.M. and Ariëns, E.J. (1962) Receptor reserve and threshold phenomena. II. Theories on drug-action and a quantitative approach to spare receptors and threshold values. Arch. Int. Pharmacodyn. 136:385-413.

Whaley, B.S., Yuan, N., Birnbaumer, L., Clark, R.B. and Barber, R. (1994) Differential expression of the beta-adrenergic receptor modifies agonist stimulation of adenylyl cyclase: a quantitative evaluation. Mol. Pharmacol. 45(3):481-489.

2. CHARACTERIZATION OF RECEPTORS BASED ON RECEPTOR-MEDIATED RESPONSES

The previous chapter described the evolution of the receptor concept and the early appreciation for the complexity that can exist between receptor occupancy and the ultimate physiological response. The present chapter summarizes methods used to characterize the specificity of a receptor-elicited response and strategies for determining the affinity constants of agonist, partial agonist, antagonist and inverse agonist agents at these receptors based on measurements of functional response.

CHARACTERIZATION OF RECEPTOR SPECIFICITY

The analytical methods described had their origin in studies of receptor-mediated responses in native tissue. These same analyses, however, are useful for analysis of response in less complex preparations and, in fact, can even be applied to cell-based assays of heterologous receptor function in high throughput drug screens.

The very existence of receptors was predicted from numerous observations demonstrating the extraordinary specificity with which a response is elicited or antagonized when a series of drug homologs is evaluated in biological preparations. Consequently, it is the *specificity* of a drug or hormone action that persuades the investigator that an observed effect is receptor-mediated rather than a nonspecific phenomenon independent of specific ligand-receptor interactions.

For endogenous stimuli or agonist drugs, the specificity of the putative receptor has been evaluated classically by determining the order of potency of a series of analogs in eliciting the desired response, e.g., contraction, secretion, ion or nutrient transport. Specificity is more easily demonstrated for small molecules than polypeptide hormones, because it is easier to prepare congeners with incremental modifications in substituent groups for small molecules. A classic example of using **order of agonist potency** in pinpointing a specific receptor's role is the pioneering work of Raymond Ahlquist (1948). Ahlquist demonstrated the existence of two receptor populations, now called α- and β-adrenergic receptors, that mediate the physiological effects of the native catecholamines, epinephrine and norepinephrine. Ahlquist observed that catecholamines evoked smooth muscle contraction with an order of potency of norepinephrine > epinephrine > isoproterenol, and he termed these effects "alpha" (α). In contrast, Ahlquist noted that increases in cardiac chronotropy (rate) and inotropy (contraction) as well as smooth muscle relaxation were elicited by catecholamines with an order of potency of isoproterenol > epinephrine > norepinephrine. Ahlquist attributed these latter effects of catecholamines to a distinct population of adrenergic receptors, which he termed "beta" (β). Thus, differing orders of agonist potency provided the original evidence for the existence of two adrenergic receptor populations, and remains the most common means for assessing receptor specificity today. It cannot be emphasized enough, however, that this method for comparing the relative potency of agonists is only useful if the agonists being studied possess the same efficacy, which often is not tested independently (see Furchgott [1972]; Kenakin [1987a]). An example of how agents with differing efficacies in different tissues can erroneously suggest the existence of distinct receptor populations is the reversal of the order of potency of oxymetazoline versus norepinephrine in eliciting α_1-adrenergic receptor contractions in rat anococcygeus muscle compared to rat vas deferens. Because oxymetazoline is a partial agonist and these two tissues have a profound difference in "spare receptors" (i.e., relationship between occupancy and response), the order of potency of these two agents is reversed in these two tissues (Kenakin [1984b]). This example emphasizes that although the relative potency of agonists and partial agonists can be compared in a given tissue, it is unlikely that the dose-ratios for a pair of agonists in eliciting a particular response will be similar from tissue to tissue, since efficacy may vary from one target tissue to another due to differences in "receptor reserves" (i.e. receptor density, effector molecules and other modulators of the occupancy-response relationship).

The ability to assign multiple responses to a single, known receptor (encoded by a heterologously expressed cDNA) has revealed that the order of potency of agonists, even when interacting with a single receptor population, can vary for different signal outputs if these agonists elicit or stabilize receptor conformations that have differing efficiencies in coupling to or activating these diverse signal outputs, e.g. GTPγ^{35}S binding, IP_3 production, Ca^{2+} mobilization, secretion, and contraction (e.g. Berg et al. [1998]). These findings are troublesome for investigators trying to define the "specificity" of a receptor. Although differing orders of agonist potency for summated responses (e.g. contraction, secretion) in native tissues was classically *de facto* evidence for involvement of different receptors in eliciting response, studies with isolated cDNAs for a single receptor but measuring different signal outputs and/or monitoring agonists of differing efficacies for activating these outputs has revealed the inherent complexity in defining receptor specificity using agonist agents alone.

A second criterion of a specific receptor-mediated event is the **selectivity of blockade by antagonist agents**. For example, Ahlquist's insightful proposal that distinct α- and β-adrenergic receptors mediate catecholamine action was corroborated by later observations that β-adrenergic effects were selectively blocked by dichlorisoproterenol (later appreciated to be a partial agonist) and propranolol, whereas α-adrenergic effects were selectively antagonized by phentolamine and phenoxybenzamine. Similarly, the subsequent subdivision of β-adrenergic receptors into β_1- and β_2-adrenergic receptor subtypes and α-adrenergic receptors into α_1- and α_2-adrenergic receptor subtypes was based primarily on the selectivity of different antagonists in blocking catecholamine effects in a variety of tissues (Schild [1973]; Berthelson and Pettinger [1977]). Unlike for agonists, the order of antagonist potency should be characteristic of a particular receptor whatever tissue preparation is employed. As discussed later, affinity constants for competitive antagonists, i.e. *null* antagonists with no intrinsic inverse agonist activity, are readily measurable (in contrast to those for agonists, partial agonists, and inverse agonists), and these constants can be helpful in classifying receptors. Based on receptor theory, it is expected that when different agonists interact with the same receptor population, the affinity constant calculated for a pure competitive antagonist should be the same regardless of which agonist is used to provoke the measured response. An assessment of whether the same K_I (or K_{D_B}) for an antagonist is obtained in the presence of several agonists provides insight into which agonists converge on a common receptor population that also is recognized by the antagonist.

There are limitations, however, in concluding that two agents act via an identical receptor population if they mutually antagonize one another's physiological responses. Thus, counter-regulatory effects mediated via distinct receptor populations are a fundamental mechanism by which a physiological steady state is maintained. These counter-regulatory effects represent *functional* antagonism, although the agents involved elicit their effects through distinct receptors rather than by competing for occupancy of the same receptor population. For example, β-adrenergic effects on cardiac inotropy and chronotropy continually are countered by acetylcholine acting via muscarinic receptors. Catecholamines elicit their effects on cardiac function by elevating intracellular cAMP levels and regulating Ca^{2+} currents, whereas muscarinic agents decrease cAMP levels, suppress voltage-gated Ca^{2+} currents and activate hyperpolarizing K^{+} currents. If one tests the effects of acetylcholine on isoproterenol-stimulated cAMP accumulation in cardiac tissue, one would observe a concentration-dependent rightward shift of the isoproterenol concentration-response curve when acetylcholine is added to the incubation. This apparent competitive antagonism might lead the naive observer to conclude that acetylcholine is a β-adrenergic antagonist and counters the effects of epinephrine by competing for agonist binding at the β-adrenergic receptor recognition site. One line of evidence that confirms that isoproterenol and acetylcholine elicit opposing effects on signaling pathways via independent populations of receptors is the observation that propranolol blocks the effects of isoproterenol, but not those of acetylcholine, on cardiac cells. Conversely, the muscarinic antagonist atropine blocks the effects of acetylcholine but not those of isoproterenol on this system.

Two independent experimental approaches beyond order of agonist and antagonist potency have been useful for differentiating the receptor(s) involved in mediating particular biological effects: (1) studies of protection against irreversible receptor blockade by reversible agonists or antagonists, and (2) cross-desensitization experiments. These methods were particularly important in early characterizations of receptor properties in native tissues, and also rely on the specificity of the receptor in interacting with particular agonist and antagonist agents.

To exploit the strategy of **protection against irreversible receptor blockade** requires the availability of an irreversible agent that reacts chemically with the same receptor recognition site as does the agonist (or antagonist), and thereby inhibits receptor-mediated functions by decreasing the density of available receptors and not by modification of some other domain of the receptor molecule or by interfering with receptor-effector coupling. As a result of binding to the receptor site, the irreversible agent causes a persistent blockade of the receptor over the time-course of the experiment. If the irreversible agent is incubated with the test tissue in the presence of a reversible agonist or antagonist that interacts with the same recognition site(s) as the irreversible ligand, then the *rate* of receptor inactivation by the irreversible agent will be slowed by competition of the reversible and irreversible agents for receptor occupancy. In contrast, when particular reversible agonists or antagonists do not afford protection, the data are consistent with the interpretation that these agents do not interact with the binding site modified by the irreversible antagonist. To assess whether a series of drugs can protect against receptor inactivation, a target tissue or cell preparation is incubated with an irreversible agent in the absence or presence of reversible agents for varying periods of time, after which the incubation is terminated by extensively washing the biological preparation. The extent of receptor inactivation that occurred during incubation with the irreversible agent is assessed by determining the extent to which an agonist can still elicit its characteristic effect in the treated preparation when compared with control preparations. In these studies, a control incubation with the protectant and *no* irreversible agent must be performed to permit assessment of whether the washing protocol used to terminate the incubation was sufficient to remove all of the protecting drug from the bathing medium, and of sufficient duration to permit dissociation of reversibly bound ligand from the tissue receptors. The most convincing evidence that reversible agonists or antagonists are interacting with (and thus protecting) the same receptor site inactivated by the irreversible ligand is that the presence of the reversible ligand decreases the *rate* of irreversible inactivation. Occasionally, when concentrations of reversible and irreversible agents are chosen appropriately, protection of the receptor by reversible agents may be apparent even at the longest interval of incubation with the irreversible antagonist. It should be remembered, however, that once the irreversible ligand occupies the receptor, the receptor binding site is inactivated and no longer vacant for occupancy by the protectant. Therefore, as the duration of incubation with the irreversible antagonist increases, the ability to detect protection against inactivation will decrease.

An example of the successful use of the protection approach to identify multiple receptor populations is a series of experiments performed by Furchgott (1954) using the β-haloalkylamine dibenamine as an irreversible antagonist. Furchgott demonstrated the probable existence of at least four independent receptor populations that evoked smooth muscle contraction in rabbit aortic strips. He conceptualized the protection experiments as either "self-protection" or "cross-protection." In self-protection experiments, the agonist present during incubation with the irreversible antagonist was the same agonist with which he subsequently assessed contraction. In "cross-protection" protocols, the agonist present during the receptor inactivation phase was different from that used to elicit contraction after extensive tissue washing. By definition, when reversible antagonists were used to protect receptors against irreversible blockade, the experimental design was one of "cross-protection." The only difference in the experimental protocol using agonists versus antagonists as the protectant is that the characteristically slower rate of antagonist dissociation from receptors requires a longer duration after washout of the protecting antagonist before retesting the agonist-elicited response. Using cross-protection studies, Furchgott demonstrated that cross-protection occurred among norepinephrine, epinephrine and isoproterenol; he used this as evidence to conclude that these three agonists all acted on the same receptor, later defined as the α-adrenergic receptor. By contrast, none of these catecholamines could afford cross-protection against inactivation of receptors for histamine, acetylcholine, or serotonin, and none of these latter agents protected among themselves or against inactivation of the catecholamine binding site. The above findings were taken together as evidence of the existence of distinct receptors for histamine, acetylcholine, and serotonin (in addition to those for catecholamines) that could mediate contraction of the rabbit aorta.

Protection against irreversible inactivation as a strategy for delineating multiple populations (or not) of receptors has limitations beyond the need for highly specific agents to serve as protectants. Thus, the existence of spare receptors in a tissue preparation also can give rise to confounding results. Even after a major fraction of a particular receptor population is inactivated, a high concentration of agonist may still elicit a full physiological response. This might lead an investigator to the erroneous conclusion that the irreversible antagonist was not interacting with the particular receptor under study. However, this potential limitation can be overcome by comparing the dose-response relationship for the agonist before and after multiple treatments with the irreversible antagonist that block increasing fractions of the putative receptor population. When a generous receptor reserve exists, irreversible receptor blockade of the "spare" receptors will first result in an increase in the EC_{50} for the agonist but no decline in maximal response, whereas progressive inactivation of the receptor population will ultimately result in a further increase in the EC_{50} *and* a decrease in the maximal response elicited by the agonist (cf. figure 2-1A).

A final experimental approach that has been used to delineate the specificity of the receptor population involved in a particular physiological response is that of **cross-desensitization** (Schild [1973]). Prolonged exposure to an agonist often results in a decline in the maximal response that can be elicited by that agonist. This agonist-induced decline in response has been referred to as *tachyphylaxis* or *desensitzation*, and cross-desensitization studies exploit this property of agonists. Thus, if exposure to agonist A results in a decline in subsequent sensitivity to agonist A as well as to agonist B, but not to agonist C, then one interpretation of these findings is that A and B interact with a common receptor and C interacts with a distinct receptor population(s). This approach has been used successfully to demonstrate a multiplicity of functional receptors for opiates and opiate-mimicking peptides in the central nervous system (see Schultz et al. [1980]). There is a serious limitation to this approach, however, in that it assumes for its interpretation that the agonist utilized elicits only a "homologous" desensitization. Homologous desensitization occurs when an agonist interferes *only* with physiological processes elicited by the particular receptor population with which that agonist interacts. In contrast, "heterologous" desensitization results when an agonist can desensitize a physiological response to subsequent stimulation by not only its own receptor but also by distinct receptor populations that activate the same response. For example, let us assume that α-adrenergic, muscarinic cholinergic, and serotonergic receptors all elicit secretion of stored contents from a particular target tissue. If this tissue is exposed to an α-adrenergic agonist for a prolonged period of time and becomes refractory to the addition of fresh agonist to the incubation, it has become desensitized. If addition of acetylcholine or serotonin still elicits the same extent of secretion characteristic of these agents in fresh tissue, then the catechoiamine-induced desensitization is said to be homologous, i.e., it only affects cellular activation via the α-adrenergic receptor. If incubation with an α-adrenergic agonist renders the tissue insensitive to acetylcholine and to serotonin as well, then the desensitization evoked by this agent is termed heterologous, as it affects not only α-adrenergic but also muscarinic and serotonergic activation of the target cell, suggesting that molecular events downstream of the α-adrenergic recptor, and shared by the muscarinic and serotonergic receptors, are involved in the desensitization process. Consequently, cross-desensitization experiments are only interpretable if the agonist used to induce tachyphylaxis evokes a homologous desensitization of the target tissue.

In summary, it is clear that each of the four experimental approaches outlined above for determining receptor specificity in native tissues is of potential value in assigning the specificity of a response to a particular receptor population. Because of the inherent limitations in each approach, however, the most definitive conclusions will result from combining some or all of these lines of experimental evidence. With the advent of molecular cloning, the specificity might be readily defined by expression of a single cDNA clone and defining specificity for a discrete signal transduction output, or comparing specificity for multiple outputs (Berget et al. [1998]). However, the existence of multi-subunit receptors, even for GPCRs, can sometimes confound the interpretations of even these seemingly straightforward experiments (Kuwasako et al. [2004]).

DETERMINING EQUILIBRIUM DISSOCIATION CONSTANTS (K_D VALUES) FOR RECEPTOR-LIGAND INTERACTIONS BASED ON MEASUREMENTS OF RECEPTOR-MEDIATED RESPONSE

The determination of K_D values for receptor-ligand interactions from receptor-mediated response data involves multiple assumptions, or requirements of experimental systems. These assumptions were originally outlined by Furchgott (1966):

1. The response (e.g. of a tissue) should be due solely to the interaction of a hormone, neurotransmitter, or agonist drug with one type of receptor, and should not be a composite reflection of the effects of two receptor populations or secondary effects of an agonist, such as agonist-provoked neurotransmitter release or agonist-induced changes in blood flow, that might occur in an intact tissue preparation.
2. The altered sensitivity to an agonist observed in the presence of a competitive antagonist should be solely a result of competition between an agonist and antagonist for a shared recognition site.
3. The response obtained following addition of a given concentration of agonist should be measured at a time when the maximal response which that concentration of agonist can elicit has been reached. Similarly, allowing desensitization (or "fade") to occur will result in an underestimation of agonist potency. Biological preparations especially suitable for determination of the K_D values for receptor-ligand interactions are those that maintain a maximal level of response for a reasonable length of time, and do not manifest time-dependent desensitization or sensitization of the response. The experimental design always should include proper controls to permit measurement of, and thus correction for, any changes in sensitivity (e.g., desensitization or sensitization) to agonist during the time-course of the experiment.
4. When agonists or competitive antagonists are added to the incubation, the concentration of ligand free in solution should be maintained at a known level. Losses due to drug uptake or to chemical or enzymatic degradation of the ligand must either be prevented or overcome by continual re-addition of the appropriate concentration of ligand.

Determination of K_D Values for Receptor-Agonist Interactions, K_{D_A}

As might be expected, determining the equilibrium dissociation constant (K_D) for binding of an agonist to its receptor using biological response data is not a straightforward procedure. If the assumptions of A. J. Clark were correct and tissue response was directly proportional to the fraction of receptors occupied, then the concentration of a hormone or an agonist drug that elicited a half-maximal response under steady-state conditions would be a direct measure of the K_D for agonist for formation of the agonist-receptor complex.[1] However, as described in chapter 1 (cf. figure 1-2) and emphasized in earlier sections of this chapter, maximal receptor-mediated effects can be elicited as a consequence of occupancy of only a small fraction of a total receptor population. Thus, the EC_{50} (concentration eliciting a half-maximal effect) for eliciting the biological response is often less than the K_D value for agonist-receptor interactions. As indicated in chapter 1, Stephenson formalized this nonlinear relationship between receptor occupancy and biological response by stating that the response of a tissue is some undefined function of stimulus S. He defined S as the product of the efficacy of an agonist, e, times the fraction of receptors occupied by the agonist:

$$\text{Response} = f(S) = fe\left(\frac{[RA]}{[R]_{TOT}}\right)$$

Based on the fundamental principles introduced by Stephenson, both Stephenson (1956) and Furchgott (1966, 1972) developed a method for determining the K_D value for agonists, K_{D_A}. Dose-response data for a particular agonist before and after irreversible receptor blockade are obtained. For certain receptor populations, a reasonably well-characterized drug may be available for this purpose, but for others an irreversible drug may not be

[1] This textbook will maintain a uniformity of nomenclature throughout chapters 2–6. Although many originators of the algebraic relationships between concentration of agonist added and receptor occupancy attained refer to K_A as the equilibrium dissociation constant for agonist, this term is confusing, as the same term commonly is used to refer to equilibrium association constants, in units of M^{-1}. The term K_D always will be used to refer to equilibrium dissociation constants, in molar units. The K_D value for agonists will be denoted as K_{D_A}, for partial agonists as K_{D_P}, for antagonists as K_{D_B} and inverse agonists as K_{D_I}. There are several other examples where symbols used in original articles are changed in the mathematical descriptions summarized in the text, to emphasize the shared concepts inherent in the analyses and to minimize confusion caused by idiosyncratic nomenclature.

available. Crosslinking an antagonist ligand to a receptor using a bifunctional reagent might mimic the properties of a covalent ligand or, alternatively, blockade by an antibody directed against the binding domain(s) of the receptor might offer an alternative approach for incrementally decreasing the population of functional receptors on the cell surface. It may also be possible to inactivate receptor populations incrementally and irreversibly using reagents that covalently modify amino acid side chains. For example, several receptor populations appear to be sensitive to alkylation with N-ethylmaleimide, a sulfhydryl residue-directed reagent, or to reagents that modify the ϵ-NH_2 group of lysine. Finally, it has been demonstrated that the equivalent of irreversible receptor blockade, i.e., irreversible receptor inactivation, can be accomplished by exposure of intact cells (Kono and Barham [1971]) or intact tissue (Lin and Musacchio [1983]) to proteases under conditions that do not alter the subsequent viability of the biological preparation.

Whether general or selective agents are employed to effect irreversible receptor blockade, the primary criterion for the validity of the procedure is that treatment must perturb only the receptor and not alter in any way the response system or receptor coupling to it. Once a method is in hand to decrease incrementally the concentration of receptor binding sites, agonist dose-response curves are obtained from control preparations and preparations in which receptors have been inactivated incrementally. For agonists possessing high efficacy (thus requiring occupancy of only a small fraction of the receptor population to elicit a maximal response), data obtained resemble those shown schematically in figure 2-1A. Thus, as an increasing fraction of the receptor population is inactivated, the agonist dose-response curve shifts to the right. Ultimately, the decrease in available receptors is sufficient to cause a decrease in the maximal response. For agonists possessing a lower efficacy (meaning that a larger fraction of the total receptor population must be occupied to elicit the maximal response), a shift to the right in the agonist dose-response curve may be subtle or not observed at all but may instead result in an immediately detectable decrease in the maximal response that can be elicited by the agonist.

The only assumption on which further analysis of the data shown in figure 2-1 is based is that an equal number of agonist-receptor complexes elicit equal responses, both before and after irreversible receptor blockade, i.e. the reagent used to effect receptor blockade does not also perturb the relationship between occupancy and response.

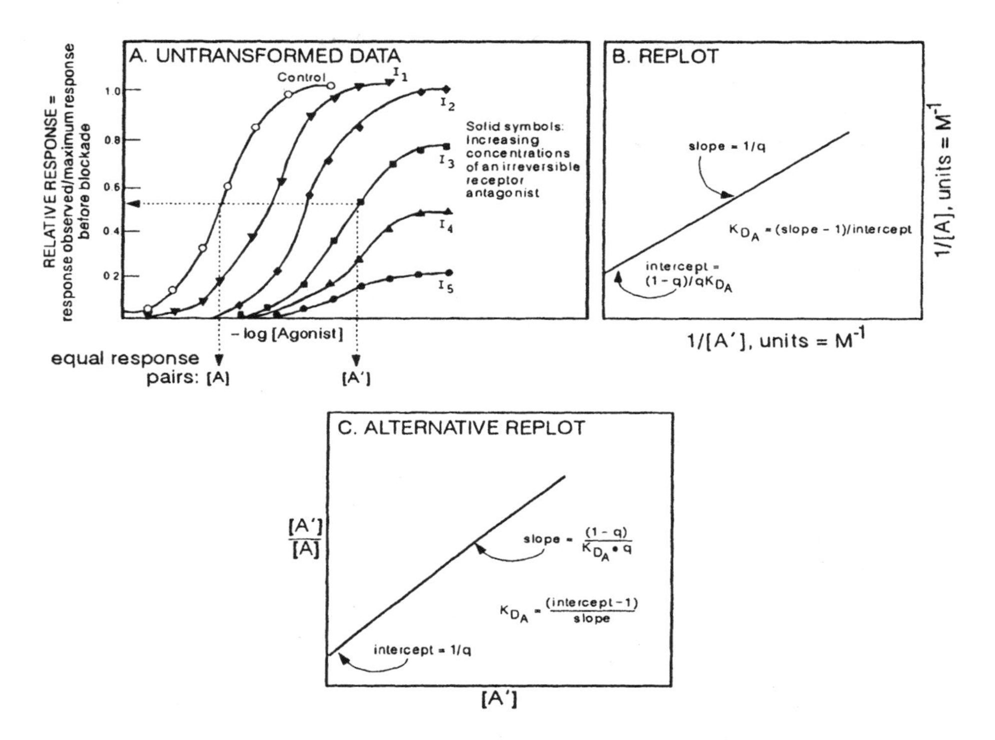

Figure 2-1. Determination of the K_D for receptor-agonist interactions (K_{D_A}) utilizing the technique of irreversible receptor blockade. Dose-response relationships are determined before (control) and after exposure of tissue preparations to increasing concentrations (A) of an irreversible antagonist, I_n . To obtain values for K_{D_A}, data can be replotted according to equation 2.7 (B), where q is the fraction of the receptor population remaining following irreversible receptor blockade, or according to equation 2.7A (C).

The algebraic relationships that result in the ability to determine the K_{D_A} for receptor-agonist interactions using experimental findings such as those shown in figure 2-1B are the following:

$$\frac{\text{observed response}}{\text{maximal response}} = \frac{E_A}{E_{\max}} = f(S) = fe\left\{\frac{[RA]}{[R]_{TOT}}\right\} \tag{2.1}$$

The above equation is a restatement of the basic concept of Stephenson. However, Furchgott found it useful to modify this formula by introducing the term $\in$, the **intrinsic efficacy**, where $e = \in[R]_{TOT}$. Therefore, substituting $e = \in[R]_{TOT}$ into equation 2.1 yields:

$$\frac{E_A}{E_{\max}} = f \in \left\{ [R]_{TOT} \cdot \frac{[RA]}{[R]_{TOT}} \right\} = f \in \{[RA]\} \tag{2.2}$$

By introducing the term $\in$, which has dimensions of the reciprocal of receptor concentration, Furchgott resolved the efficacy term of Stephenson (e) into two components: the drug-dependent component $\in$ and the tissue-dependent component, $[R]_{TOT}$. This resolution of the two components inherent in drug efficacy emphasizes that Stephenson's efficacy term e is dependent on the total concentration of available, functional receptors, $[R]_{TOT}$, and that two different biological preparations containing the same response system will have different values for e depending on the extent to which $[R]_{TOT}$ differs in the two biological preparations. This definition of $e = \in [R]_{TOT}$ and its substitution into the equations, however, does not affect the ultimate determination of K_{D_A} values.

Mass action relationships for a biomolecular interaction dictate that:

$$\frac{[RA]}{[R]_{TOT}} = \frac{[A]}{K_{D_A} + [A]} \tag{2.3}$$

Substitution into equations 2.1 and 2.2 yields:

$$\frac{E_A}{E_{\max}} = f(S) = fe\left\{ \frac{[A]}{K_{D_A} + [A]} \right\} = f \in \left\{ [R]_{TOT} \ \frac{[A]}{K_{D_A} + [A]} \right\} \tag{2.4}$$

If the treatment of a biological preparation with an irreversible antagonist reduces the concentration of total active receptors, $[R]_{TOT}$, to a fraction, q, of the original $[R]_{TOT}$, *then* the effective efficacy becomes qe or $q\in [R]_{TOT}$, and

$$\frac{E'_A}{E'_{\max}} = (S') = fe\left\{ q \ \frac{[A']}{K_{D_A} + [A']} \right\} = f \in \left\{ q[R]_{TOT} \frac{[A']}{K_{D_A} + [A']} \right\} \tag{2.5}$$

where E'_A, $E'_{\max}$, S' and $[A']$ correspond to E_A, $E_{\max}$, S and $[A]$, respectively, following irreversible inactivation of $[R]_{TOT}$ to $q[R]_{TOT}$. When comparing *equal responses*, i.e., when stimulus S before receptor inactivation with an irreversible antagonist is assumed to be equal to the stimulus S' after inactivation, then $S = S'$ and, therefore,

$$\in [R]_{TOT} \cdot \frac{[A]}{K_{D_A} + [A]} = \in q[R]_{TOT} \cdot \frac{[A']}{K_{D_A} + [A']}$$

eliminating $\in \cdot [R]_{TOT}$ yields

$$\frac{[A]}{K_{D_A} + [A]} = \frac{q[A']}{K_{D_A} + [A']} \tag{2.6}$$

The above mathematical manipulation has allowed cancellation of the receptor density term $[R]_{TOT}$ and permits an estimate of K_{D_A} based entirely on response data. Note also that the method does not require knowing a value for *f*, the function which relates occupancy to response.

Equation 2.6 can be rearranged to the form:

$$\frac{1}{[A]} = \frac{1}{q[A']} + \frac{(1-q)}{q \cdot K_{D_A}} \tag{2.7}$$

This is the linear transformation ($y = mx + b$) shown in figure 2-1B. The slope of the line of a plot of $1/[A]$ versus $1/[A']$ is equal to $1/q$ and the y intercept equals $(1 - q)/qK_{D_A}$. The K_D value for agonist, K_{D_A}, equals (slope - 1)/intercept. Furthermore, q (fraction of receptors remaining active) equals 1/slope.

Equation 2.6 can be rearranged to a linear transformation other than that shown in equation 2.7. For example, by multiplying both sides of equation 2.7 by A', one obtains:

$$\frac{[A']}{[A]} = 1/q + \frac{[A'](1-q)}{K_{D_A} \cdot q} \tag{2.8}$$

A plot of $[A']/[A]$ versus $[A']$ yields a straight line with an upward slope. Statistical software tools can be used to provide estimates of the intercept and slope, from linear regression analysis, and the error of each of these two values. This line has a y intercept equal to $1/q$ and a slope equal to $(1 - q)/K_{D_A} \cdot q$. The K_{D_A} value can be calculated as (intercept - 1)/slope. This alternative procedure for obtaining the q and the K_{D_A} value is shown in figure 2-1C. Furchgott (personal communication) favors the replot shown in figure 2-1C because data derived from studies at very low concentrations of agonist come at the beginning of the plot. Hence, these data do not significantly influence the slope and intercept of the plot in figure 2-1C to the extent that they do in the double reciprocal plot in figure 2-1B. This is desirable because if there is any discrepancy between $[A]_{added}$ and the *effective* concentration of agonist at the receptor because of tissue uptake or degradation, this discrepancy (i.e., error) would be greatest at very low concentrations of agonist.

To transform the normalized dose-response data shown in figure 2-1A to the replots shown in figures 2-1B and 2-1C, *pairs* of concentrations of agonist resulting in *equal responses* before (i.e., $[A]$) and after (i.e., $[A']$) incubation with an irreversible antagonist are compared. Effective use of this technique requires that the dose-response curve after receptor inactivation has a depressed maximal response. Naturally, it is unlikely that experimentally added concentrations of agonist before and after irreversible receptor blockade will turn out to be pairs of $[A]$ and $[A']$ that elicit equal biological responses. Consequently, a curve must be drawn between the data points obtained, necessarily introducing some error into the subsequent analysis. The data analysis (and resultant determination of K_{D_A}) is most practical and most valid statistically if the concentration pairs $[A]$ and $[A']$ are selected for replotting by starting with values of $[A']$ actually added to the preparation exposed to an irreversible antagonist and then estimating the value of $[A]$ that gives an equal response in the control preparation by inspecting the dose-response curve in the untreated tissue. The suggestion is practical, in that there is always a response (E_A) in the control preparation that corresponds to a response (E_A') in the irreversibly inactivated preparation. However, the reverse will not always be true. The statistical validity of this approach is that the replot obtained using data pairs $[A]$ and $[A']$ can be analyzed more easily by least squares fitting, since $[A']$ is determined experimentally (see Furchgott [1966], Parker and Waud [1971], and Thron [1973]). It also is worth reiterating that the analysis depicted in figure 2-1 requires only a single dose-response curve following irreversible receptor blockade for comparison with the control curve. However, confidence in the K_{D_A} value obtained is increased if the calculated value is similar when several levels of receptor inactivation are achieved as a result of incubation of the receptor-response system with various concentrations (or with a single concentration at various times) of an irreversible antagonist.

Certain assumptions were made in developing the theory for calculation of K_{D_A} following irreversible receptor blockade which must be met in the experimental protocol in order for the approach in figures 2-1B or 2-1C and the resultant analysis to be valid (Furchgott [1972]).

1. The agonist elicits its measured response by interacting with a single receptor population.

2. The interaction of agonist with the receptor is a reversible, bimolecular reaction operating according to the principles of mass action, such that $[RA]/[R]_{TOT}$ can appropriately be equated with $[A]$ $(K_{D_A} + [A])$ (cf. equation 2.3).
3. The receptor population involved is "uniform" with respect to K_D (this is inherent in the second assumption). Thus, it is assumed that neither negatively nor positively cooperative interactions occur among the receptors, so that the K_D of the receptor for agonists is independent of the degree of occupancy of the receptor population. This methodology does *not* apply if the receptor-response system is cooperative in nature (see Thron [1973]).
4. The relationship between stimulus S and concentration of agonist-receptor complexes $[RA]$ (i.e., the intrinsic efficacy, $\in$) remains the same after irreversible inactivation of a part of the receptor preparation.
5. The $[A]$ plotted as the "free drug" available to receptors is essentially equal to the concentration of drug added to the bathing solution or incubation medium.
6. Desensitization (or sensitization) to the agonist does not occur during the time-course of the experiments.
7. Inactivation of receptors is completely irreversible, such that the value of q does not change after washout of the irreversible blocker. Also, when evaluating the effects of an irreversible antagonist, the unbound antagonist should be removed by extensive washing of the preparations, so its concentration is essentially zero during subsequent measurements of response.

The development of experimental strategies to deduce the K_{D_A} for an agonist at its receptor based on agonist-elicited response data in a complex physiological system preceded methodologies for quantitating receptor density with radioligand binding or molecular insights into the biochemical events linking receptor occupancy to the ultimate physiological or pharmacological effect. However, as more molecular insights are gained about the receptor-elicited activation process, concern about the validity of the receptor inactivation method for determination of K_{D_A} has arisen (Leff et al. [1990] and references therein). For example, two tenable models for agonist activation of receptors are the **isomerization model**, e.g., for ligand-gated ion channels (del Castillo and Katz [1957]), and the **ternary complex model** and extensions of that model (cf. chapter 4) for receptors that control cellular processes via heterotrimeric GTP-binding proteins (DeLean et al. [1980]).

Subset equations of the:

Isomerization Model

$$A + R \overset{K_{D_A}}{\rightleftharpoons} AR \rightleftharpoons AR^*$$
$$\downarrow$$
Response

Ternary Complex Model

$$A + R \overset{K_{D_A}}{\rightleftharpoons} AR \rightleftharpoons ARG$$
$$\downarrow$$
Response

In the isomerization model, agonist occupancy of the receptor evokes a conformational change in the receptor to an active state (R^*), which is responsible for the receptor-elicited response. In the ternary complex model, the agonist-occupied receptor interacts with a heterotrimeric ($\alpha\beta\gamma$) G-protein (G) to facilitate GTP-binding to the α subunit and activation of subsequent α and/or $\beta\gamma$ subunit-dependent responses. In *both* cases, if either AR^* *or* ARG accumulates, then the equilibrium will be pulled to the right, and the estimate of K_{D_A} will suggest that the receptor has a higher affinity for agonist than its intrinsic affinity would dictate. It can be shown for both the ternary complex and isomerization models that the estimation of affinity constants for partial agonists is subject to less error than estimations for full agonists, since these partial agonists convert only a small fraction of AR to AR^* or ARG. Despite the theoretical concern, however, only minimal errors in estimated affinity constants appear to have been made when a comparison of K_{D_A} values obtained in tissue-bath studies to those obtained using direct radioligand binding analyses (cf. chapter 3) has been undertaken (Leff et al. [1990]). Thus, classical pharmacological theory appears to offer experimental strategies that yield reasonable estimates of receptor affinity and, at the very least, provide an initial characterization of the properties of receptors based on the responses they elicit.

Determining K_D Values for Receptor-Partial Agonist Interactions, K_{D_P}

Multiple approaches for determination of the K_D values for partial agonists have been applied in the literature; three will be described in some detail here. The first is based simply on a comparison of the dose-response curve of a partial agonist to that of a full agonist (Barlow, Scott and Stephenson [1967]; Waud [1969]). The second approach determines doses of a full agonist that are equiactive in the absence and presence of a partial agonist (Stephenson [1956]; Colquhoun [1973]). A third strategy involves irreversible receptor blockade of a sufficient fraction of the receptor population so that a partial agonist is no longer able to elicit any measurable biological response. The partial agonist is then used as a competitive antagonist of a full agonist, and its K_D for the receptor is determined as that for antagonists (Furchgott and Bursztyn [1967]). Each of these three approaches will be considered below. An alternative approach that will not be considered in detail here is that based on the model of Van Rossum and Ariëns for partial agonists (see Van Rossum [1963]), which evaluates the shift in the agonist dose-response curve caused by addition of increasing concentrations of a partial agonist to the incubation (see also Kenakin and Black [1978]).

The theory behind determining the K_D for receptor-partial agonist interactions by comparing the dose-response curves of partial agonists with those of full agonists is based on one principal assumption: that full or "strong" agonists elicit a maximal response by occupying only a small fraction of the total receptor population. Partial agonists, by definition, have a lower efficacy and must fill an appreciably greater fraction of the receptors to elicit a response. As described earlier, the observed response is some function of stimulus $S = eY_A$, where Y_A is the fraction of receptors occupied by the agonist. Fractional receptor occupancy Y_A can be expressed as:

$$Y_A = \frac{[RA]}{[R]_{TOT}} = \frac{[A]}{[A] + K_{D_A}}$$

The assumption is that for a full agonist, A, the $[A]$ resulting in a response is very small relative to its K_D for interacting with the receptor. Thus, for a full agonist A:

$$Y_A = \frac{[A]}{K_{D_A}} \tag{2.9}$$

In contrast, for a partial agonist, P,

$$Y_P = \frac{[RP]}{[R]_{TOT}} = \frac{[P]}{[P] + K_{D_P}} \tag{2.10}$$

and this relationship cannot be simplified further.

To determine experimentally the K_D for a partial agonist P, one obtains a dose-response curve for the full agonist A and the partial agonist P. Since, by definition,

$$S = e_A\, Y_A$$

$S = e_P Y_P$

If one compares the concentrations of A and P that elicit *equal responses*, then

$$e_A \frac{[A]}{K_{D_A}} = e_P \frac{[P]}{[P]+K_{D_P}} \tag{2.11}$$

and by rearranging to a linear transformation one obtains:

$$\frac{1}{[A]} = \left[\frac{e_A}{e_P} \cdot \frac{K_{D_P}}{K_{D_A}}\right]\frac{1}{[P]} + \frac{e_A}{e_P K_{D_A}} \tag{2.12}$$

Thus, a plot of $1/[A]$ versus $1/[P]$ should be linear if all assumptions regarding the ligand-receptor interactions are correct and a sufficient receptor capacity exists so that equation 2.9 is valid.

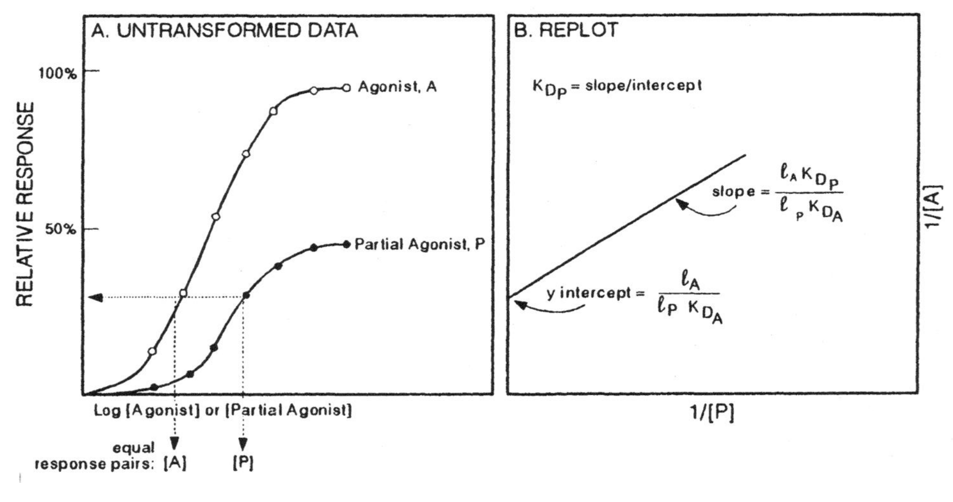

Figure 2-2. Determination of the equilibrium dissociation constant for a partial agonist, K_{D_P}, determined by comparison of dose-response curves of the partial agonist with a full agonist.

Figure 2-2A shows a schematic diagram comparing the response obtained with a full agonist and a partial agonist, and figure 2-2B shows the data transformation to obtain the equilibrium dissociation constant for the partial agonist P. The slope of the plot of $1/[A]$ versus $1/[P]$ equals

$\frac{e_A K_{D_A}}{e_P K_{D_A}}$, and the intercept on the y axis is $\frac{e_A}{e_P K_{D_A}}$.

The equilibrium dissociation constant for partial agonist P can be calculated as follows:

$$K_{D_P} = \frac{\text{slope}}{y \text{ intercept}} = \frac{\left[\frac{e_A}{e_P} \cdot \frac{K_{D_P}}{K_{D_A}}\right]}{\left[\frac{e_A}{e_P} \cdot \frac{1}{K_{D_A}}\right]} = K_{D_P}$$

Furthermore, by knowing the equilibrium dissociation constant for the agonist K_{D_A} and setting an arbitrary value for the efficacy of the full agonist (e.g., $e_A = 1$), one can obtain the efficacy of the partial agonist ($e_P < 1.0$).

It must be stressed that the assumption made in the method described above is that the concentration of agonist ([A]) that produces a maximal response is much less than the equilibrium dissociation constant for the agonist K_{D_A}. If the tissue being studied does not have a receptor reserve for a full agonist, then an error term will be introduced into the calculation, such that the procedure will alter the estimate of the value for K_{D_P}:

$$K_{D_P} = \frac{\text{slope}}{y \text{ intercept}}\left(1 - \frac{e_P}{e_A}\right)$$

It is apparent that the error diminishes to zero as the difference between the efficacy of the full and partial agonist increases (see Kenakin [1984a]).

In his initial paper which introduced his concept of efficacy, Stephenson suggested another method for estimating the K_D value for receptor-partial agonist interactions. It can be shown algebraically that an estimate of K_{D_P} can be made by determining doses of a full agonist that are equiactive in the presence ([A']) and absence ([A]) of a partial agonist P. Again, the assumption is made that the concentration of agonist eliciting a maximal response is much less than the K_D value for receptor interactions with "full" or "active" agonists. The equation that relates the equiactive doses of a full agonist in the absence and presence of a partial agonist is the following:

$$[A] = \frac{[A']}{1 + \frac{[P]}{K_{D_P}}} + \left[\frac{e_P}{e_A} \cdot \frac{[P]}{K_{D_P}} \cdot \frac{K_A}{1 + \frac{[P]}{K_{D_P}}} \right] \tag{2.13}$$

and K_{D_P} is estimated from a plot of $[A]$ versus $[A']$ by the following relationship:

$$K_{D_P} = \frac{[P] \cdot \text{slope}}{1 - \text{slope}}$$

Again, if the assumption of a considerable spare receptor capacity for the "full" or "strong" agonist is not met, an error term of $(1 - e_P/e_A)$ is introduced which diminishes to zero if $e_A >> e_P$ (see Kenakin [1984a] and references therein). An even more rigorous version of this method (Kauman and Marano [1982]) examines agonist concentration-response curves in the presence of a number of concentrations of the partial agonist, thus providing a range of partial agonist concentrations over which to estimate the K_{D_P} and utilizes slopes from a range of equiactive agonist concentration plots as:

$$\log\left(\frac{1}{\text{slope}} - 1\right) = \log[P] - \log K_{D_P} \tag{2.14}$$

A plot of $\log\left(\frac{1}{\text{slope}} - 1\right)$ as the y axis versus $[P]$, log scale, yields a linear regression whose slope should not be significantly different from unity and whose intercept estimates the K_{D_P}.

A third technique for the determination of the K_D for a partial agonist (K_{D_P}) was introduced by Furchgott and Bursztyn (1967) as an "internal check" for the value of K_{D_P} determined using the procedures described above. Incubation conditions for treatment with an irreversible antagonist are determined that still permit the effects of a strong or "full" agonist to be obtained but which inactivate a sufficient fraction of the receptor population so that the partial agonist under study no longer produces a response. The interaction of the partial agonist with its receptor can then be studied as a competitive antagonist of the full agonist. Consequently, the K_{D_P} under these circumstances can be estimated as for competitive antagonists (see following section):

$$K_{D_P} = \frac{[P]}{\text{dose ratio} - 1} \tag{2.15}$$

In some systems, partial agonists are of such low efficacy that a full agonist can still produce a further response. Experimentally, one observes an elevated baseline due to the partial agonist and a rightward shift of the control agonist-response curve, due to antagonism by the partial agonist. Dose ratios of equiactive concentrations of agonist can be estimated from the agonist-dependent region of the dose response curves. Kenakin (2004) has shown that a Schild analysis (see below) under such conditions will slightly underestimate the K_{D_P} value for the partial agonist, but the error will be minimal.

Determining the K_D Value for Receptor-Antagonist Interactions, K_{D_B}

An antagonist is any agent that blocks responses to agonist-evoked, receptor-mediated responses. Antagonists can act via the agonist-binding, so-called "orthosteric" binding pocket of the receptor, or they can suppress function as non-competitive inhibitors occupying other, allosteric sites. Orthosteric competitive antagonists, classically referred to as null antagonists, will cause rightward, parallel shifts in the agonist dose-response curve. Ultimately, the effects of the antagonist will be *surmountable* by agonist. Not all orthosteric antagonists, however, are devoid of intrinsic activity. As will be discussed in greater detail in chapter 4, some antagonists have a higher affinity for an active, rather than an inactive, state of the receptor; these agents can behave as partial agonists, particularly in receptor systems rendered constitutively active due to heterologous overexpression of a particular receptor. If antagonists have a higher affinity for the inactive state(s) of the receptor, then the antagonist will express inverse agonism, particularly in a constitutively active system. Where no agonist-independent activity occurs in a system, inverse agonists may appear to behave as simple competitive antagonists. The Schild analysis, explained below, was developed to quantitate the K_D for antagonist agents (K_{D_B}) that were simple, competitive blockers at the orthosteric binding site.

When a strictly competitive, reversible antagonist interacts with a receptor population, its ability to influence receptor occupancy by an agonist is determined both by the affinity of the receptor for the antagonist and the concentration of antagonist present. Thus, at the level of receptor occupancy, two independent equilibria are occurring for agonist A and antagonist B:

$$A + R \rightleftharpoons AR$$

$$B + R \rightleftharpoons BR$$

As a consequence of AR formation, a biological effect is elicited. In contrast, no effect is elicited as a consequence of BR formation when B is a null, competitive antagonist. Instead, fewer receptors are available for occupancy by the agonist. Competitive antagonists, therefore, suppress agonist-mediated responses by blocking access of the agonist to its specific receptor.

It was Gaddum (1937, 1943) who first formulated the relationship between the fraction of receptors occupied by an agonist as a function of the concentration agonist $[A]$, of competitive antagonist $[B]$, and of their respective equilibrium dissociation constants, K_{D_A} and K_{D_B}:

$$Y_{A'} = \frac{1}{1 + \frac{K_{D_A}}{[A']}\left(1 + [B]/K_{D_B}\right)} = \frac{[A']}{[A'] + K_{D_A}(1 + [B]/K_{D_B})} \qquad (2.16)$$

where $Y_{A'}$ is the fractional receptor occupancy by the agonist that is diminished, due to the presence of antagonist, from the fractional occupancy Y_A obtained in the absence of antagonist, namely $Y_A = [A]/([A] + K_{D_A})$. $[A']$ refers to concentration of agonist evaluated in the presence of antagonist.

As shown in figure 2-3A, the effects of competitive antagonists on agonist-provoked responses are evaluated by determining the dose-response relationship for agonist A in the absence and presence of increasing concentrations of an antagonist B. As indicated above, a series of parallel rightward curves is expected in the presence of increasing concentrations of a null, competitive, fully reversible antagonist.

Since the relationship between receptor occupancy by an agonist and the ultimate response is not necessarily linear, it is not possible to determine the K_{D_B} by fitting the data of a single agonist-dose-response curve in the presence of antagonist to equation 2.16. Instead, like methods described previously for

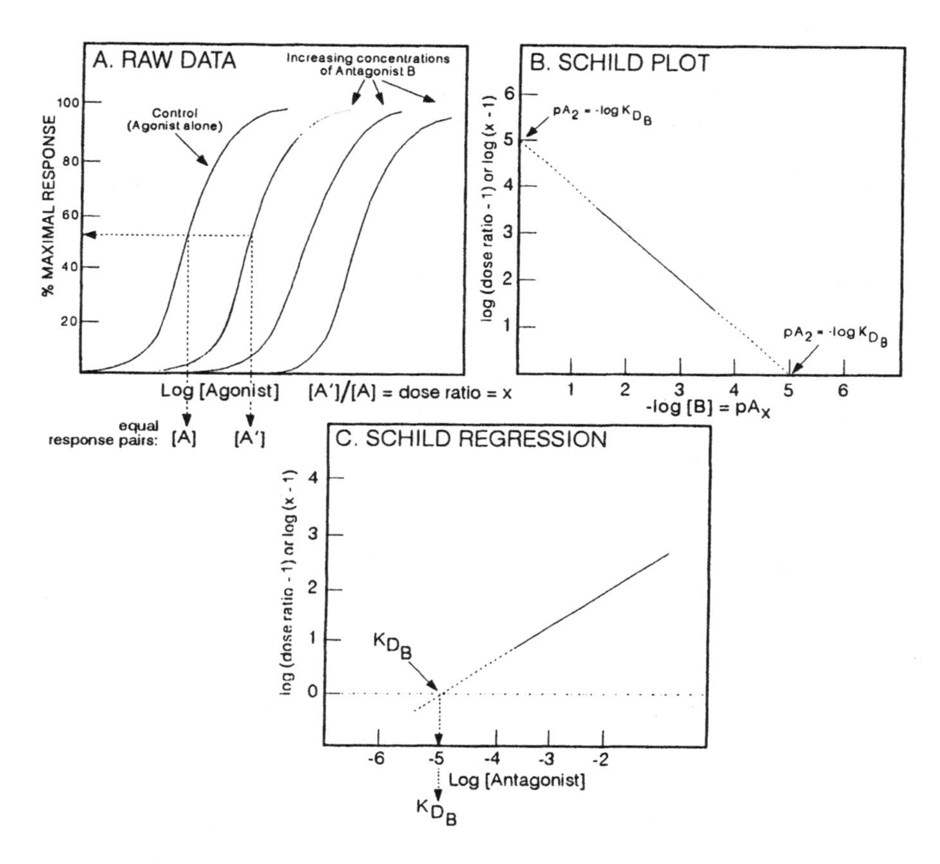

Figure 2-3. Determination of the K_D for receptor interaction with a competitive antagonist, K_{D_B}.

---

determining the affinities of receptor-agonist and receptor-partial agonist interactions, a "null method" is employed, i.e., one assumes that equal responses are elicited when an equal number of receptors are occupied by an agonist, such that

$$\frac{[A]}{[A]+K_{D_A}} = \frac{[A']}{[A']+K_{D_A}(1+[B]/K_{D_B})} \tag{2.17}$$

Schild simplified this equation to

$$\frac{[A']}{[A]} - 1 = \frac{[B]}{K_{D_B}} \tag{2.18}$$

This formulation permits the determination of K_{D_B} while making no assumptions regarding the relationship between fractional occupancy and the ultimate response. Schild denoted the ratio of agonist concentrations that elicits an equal response in the presence (A') or absence (A) of antagonists as x (Schild [1957]). This ratio is often also referred to as the *dose ratio.* Furthermore, Schild defined pA_x as the negative logarithm of the antagonist concentration, in molar units, that produced the dose ratio of x. By taking the logarithms of equation 2.18 and substituting x for $[A']/[A]$, one obtains:

$$\log (x - 1) = \log [B] - \log K_{D_B} \quad (2.19)$$

Schild's definition of pA_x as $-\log[B]$, producing a dose-ratio of x, changes equation 2.19 to:

$$\log (x - 1) = -pA_x - \log K_{D_B} \quad (2.20)$$

It can be seen that when

$$x = 2, \log (2 - 1) = \log 1 = 0$$

Thus, when

$$x = 2, pA_x = -\log K_{D_B} \quad (2.21)$$

Consequently, a concentration of antagonist that shifts the agonist dose-response curve twofold is equal to the equilibrium dissociation constant for the receptor-antagonist interaction, i.e., K_{D_B}. To determine graphically the value of pA_2 from dose-response data such as those shown schematically in figure 2-3A, Schild introduced the plot shown in figure 2-3B, i.e., $\log (x - 1)$ versus $-\log[B]$ from the linear transformation inherent in equation 2.20 (see Arunlakshana and Schild [1959]). Equation 2.20 shows that the intercept on the x axis ($y = 0$) is equal to pA_2. Furthermore, since the theory inherent in derivation of the equation dictates that the slope of this plot must equal -1, the intercept on the y axis must *also* be equal to pA_2.

The transformation of data from a plot such as that shown in figure 2-3A to the **Schild plot** in figure 2-3B is straightforward. One obtains a family of dose-response curves for the agonist: a control curve in the absence of antagonist and agonist concentration-response curves in the presence of increasing concentrations of the putative competitive antagonist. When the control curve and a curve from an antagonist-containing incubation are compared, dose ratios are obtained, i.e., a ratio of the concentrations of agonist that elicit the *same response* in the presence ($[A']$) or absence ($[A]$) of antagonist. (As mentioned above, the dose ratio = $[A']/[A]$ = x of Schild.) The dose ratio theoretically can be taken at any response level that is not near the limits of the physiological response, where experimental error is too great. However, the dose ratio most commonly is compared at the EC_{50}, i.e., the concentration of agonist that elicits 50% of the maximal response. (Refer to figure 2-3A for determination of one pair of agonist concentrations, $[A]$ and $[A']$). To obtain the best estimate of pA_2, and hence K_{D_B}, it is useful to obtain log (*dose ratio*-1) values as close to zero as possible, i.e. using concentrations of antagonist that shift the agonist dose-response curve ~2-5 fold, ensuring that the experimental data cluster around the x intercept. Furthermore, a 30-100 fold range of antagonist concentrations should be used to obtain a statistically confident estimate of the slope (Kenakin [2004]).

A number of investigators quantitate receptor-antagonist interactions using the theory inherent in the Schild analysis but a slightly different data transformation from that of figure 2-2B. This transformation is sometimes called a **Schild regression** and is shown in figure 2-3C (Schild [1947]). Recalling equation 2.19: $\log(x - 1) = \log[B] - \log K_{D_B}$, the data can also be plotted as $\log(x - 1)$ or the equivalent log (*dose ratio* - 1) on the ordinate versus $[B]$, in log units, on the abscissa. The only difference in the plot is that there is a reverse sign of the slope, i.e., simple competitive antagonism should result in a straight line of slope = 1. The intercept on the x axis of a plot of log (dose ratio - 1) versus $\log[B]$ yields a direct estimate of the equilibrium dissociation constant for the antagonist, K_{D_B}.

The linear relationship between dose ratio and K_{D_B} results from the assumption that the antagonist interacts with a homogenous population of receptors at the orthosteric site for agonists with a single, unchanging affinity. Consequently, a Schild analysis yielding a straight line with a slope of 1 is consistent with the conclusion that the antagonist under study competitively antagonizes agonist occupancy of a homogenous population of receptors whose interactions with ligand is without positive or negative cooperativity. Since receptor affinity for the antagonist should be an unchanging parameter, the K_{D_B} value calculated for the antagonist should be independent of the agonist used, and Schild plots for multiple agonists studied in the presence of the same antagonist should yield indistinguishable K_{D_B} values.

Although the graphical method of Schild is straightforward, its application to raw data may nonetheless be complicated for either experimental or biological reasons. For example, if the antagonist-induced rightward shift of the agonist dose-response curve is not exactly parallel, then the points will not lead to a good linear fit (by non-weighted least squares linear regression analysis) and the slope of the best line may not equal -1 (figure 2-3B) or 1 (figure 2-3C). Interpretations of data when the slope $\neq 1.0$

are given below. Furthermore, if there is considerable scatter around the line of slope = -1 or 1, the standard error of the intercept will be large, and there will be a corresponding large error in the estimate of K_{D_B}. The error of the extrapolation also depends on the distribution of data points: are they well spread along the line from its intercept at the y axis to its intercept at the x axis, or are the points clustered at one part of the line? The variability inherent in a poorly defined intercept (as is obtained with clustered data points) is immediately apparent when the pA_2 value calculated from the x intercept differs significantly from that calculated at the y intercept of a Schild plot (figure 2-3B).

Several statistical approaches are available for analysis of Schild plots whose slope values are precisely -1.0 or 1.0 (see Parker and Waud [1971]; Mackay [1966;1978;1982]; Tallarida et al. [1979]). Perhaps the simplest approach to assure a reasonably reliable estimate of pA_2 is to constrain the fit of the line so that the slope does equal -1 (figure 2-3B) or 1 (figure 2-3C), as theory dictates it should, and determine the pA_2 value from the line best fitted by this slope value. This approach, however, is only tenable if the experimental data generate a line whose derived slope is not significantly different from 1.0, as determined by linear regression.

Schild plots that do not have a slope of 1 may reflect more complex receptor-antagonist interactions or, perhaps, technical limitations. Inadequate antagonist equilibration times prior to the study of agonist dose-response curves also will result in nonlinear Schild plots with portions possessing slopes > 1 (Kenakin [2004]). It can take considerable time for agonist to equilibrate with receptors in the presence of antagonist, due to the characteristically slow dissociation rate of antagonists (Kenakin [2004], figure 6.14 and accompanying text). Similarly, if an uptake or enzymatic removal process is occurring such that the $[A]$ added to the incubation is not the $[A]$ available for receptor occupancy, then the *control* agonist concentration response curve will be shifted to the right of the true curve. Furthermore, if the antagonist blocks this uptake process at certain concentrations, the ultimate effect of these "side reactions" will be the appearance of even more complex Schild plots. The influence of agonist-uptake processes on Schild analysis has been dealt with quantitatively by Furchgott (1972) and Kenakin (1984a). Nonlinear Schild analyses also can arise as a result of any antagonist-produced effect independent of receptor blockade that potentiates response to the agonist. For example, if an antagonist not only blocks receptor occupancy by agonist but also results in the release of endogenous agonist, nonlinear Schild plots will result. Similarly, if an agonist is linked to stimulation of cAMP accumulation and the receptor antagonist not only blocks the receptor but also inhibits phosphodiesterase activity, the amplification of response due to the blockade of cAMP hydrolysis will result in nonlinear Schild plots. The limitation of Schild analysis for studying antagonists with more than one site of function can be addressed using "resultant plots," which are described later in this chapter.

The ability of Schild plots to reveal biological complexity, once technical contributors to non-linear Schild plots or plots with slopes $\neq$ 1.0 have been excluded, is of considerable value. Regressions with slope values < 1.0 are characteristic of receptor heterogeneity. When agonist-dependent Schild plots provide evidence for the existence of receptor heterogeneity, such as receptor subtypes (cf. Furchgott [1978] and Kenakin [1987b]), the apparent receptor heterogeneity manifested by Schild analysis will be influenced by the relative concentration of receptor populations, the relative affinity of these populations for various agonists and antagonists, and by the effectiveness of coupling of these receptor populations to the measured response. Furthermore, the ratio of the concentrations of various receptor populations (i.e., $[R]_{TOT}$) and thus agonist efficacy at these receptors ($e = \in [R]_{TOT}$) probably will vary from tissue to tissue. Consequently, the pA_2 values obtained when receptor heterogeneity exists are *not* equivalent to values for K_{D_B}, but represent a composite of all of the above influences on apparent antagonist potency.

Although Colquhoun (1973) has shown that Schild analysis can be appropriate for various cooperative models of ligand-receptor interactions, important exceptions exist where Schild analysis cannot give a precise estimate of K_{D_B}; e.g., when two molecules of agonist must bind to two cooperatively linked sites for receptor activation to occur (Sine and Taylor [1981]). An example is the nicotinic cholinergic receptor on skeletal muscle linked to Na^+ channel opening. In this case, the pA_2 value calculated from a Schild analysis does not correspond to the K_{D_B} for receptor-antagonist interactions.

Determining the Equilibrium Dissociation Constant for Inverse Agonists

When biological systems lack a "basal" or agonist-independent receptor activity, inverse agonists will produce parallel-rightward shifts in agonist-dose-response curves, i.e. will appear as null, competitive antagonists. In these situations, the K_D for the inverse agonist can be determined using Schild analysis, as described above. This is in fact why the inverse agonist properties of many antagonist drugs were not appreciated in native biological tissues or other preparations where agonist-independent, or constitutive, activity was subtle or non-existent.

Heterologous overexpression of cloned receptors, however, often results in spontaneous, agonist-independent activity of these receptors and an elevation of the baseline signaling response. Antagonists with negative intrinsic activity do not correspond to the assumptions inherent in the Schild analysis of antagonist effects on agonist-elicited responses, i.e. there will be non-parallel shifts in the dose-response curves as "basal" or constitutively active receptor signal is reduced from baseline simultaneous with a competition for agonist-evoked response. Inverse agonists, by definition, will have a higher affinity for the inactive state of a receptor (cf. two- or more state models in chapter 1 and the extended ternary complex model for G protein-coupled receptors in chapter 4; Schutz and Freissmuth [1992]; Lefkowitz et al. [1993]). This is manifested by a dose-dependent decrease in the "basal" receptor activity by inverse agonists. Kenakin (2004) has shown that the concentration of inverse agonist that decreases basal receptor activity by 50% often approximates the concentration that produces a two-fold shift to the right of the agonist dose-response curve. Intuitively, however, the ability to estimate the K_{D_B} value for an inverse agonist from its impact on basal or on agonist-evoked receptor responses will be perturbed the most for agents that have profound differences in affinity for the active versus inactive states of the receptor. Further discrepancies between EC_{50} values for decreasing basal activity or for causing two-fold rightward shifts in agonist dose-response curves and the true K_{D_B} value for inverse agonists occur in highly coupled occupancy-response systems, again characteristic of heterologous overexpression systems. The observed EC_{50} will, in fact, be greater than the K_{D_B} in highly coupled systems.

PHARMACOLOGIC RESULTANT ANALYSIS

Black (1986) initially introduced the method of pharmacologic resultant analysis to allow an estimation of the K_{D_B} for an antagonist that is known to have additional non-receptor effects, e.g. an antagonist that both blocks a receptor and, at higher concentrations, blocks neurotransmitter transport so that it alters neurotransmitter availability in the same preparation. However, as a general tool, performing resultant analysis provides further insight into the mechanism of action of an antagonist, i.e. whether it is acting as a pure competitive antagonist at the agonist-binding site or behaving as a modifier of agonist action via additional mechanisms.

Resultant analysis compares the blockade of agonist responses by two antagonists, one defined as the "reference antagonist," a known competitive antagonist, and the "test antagonist," which may have additional effects in addition to or independent of competitive antagonism. This method is similar to the additive dose ratio method of Paton and Rang (1965) to determine competitiveness, but is superior because it compensates for any secondary effects of the test antagonist by determining both the "control" dose-response curve and the curve for evaluating the reference antagonist in the presence of the test antagonist, thus nullifying any non-receptor effects of the test antagonist.

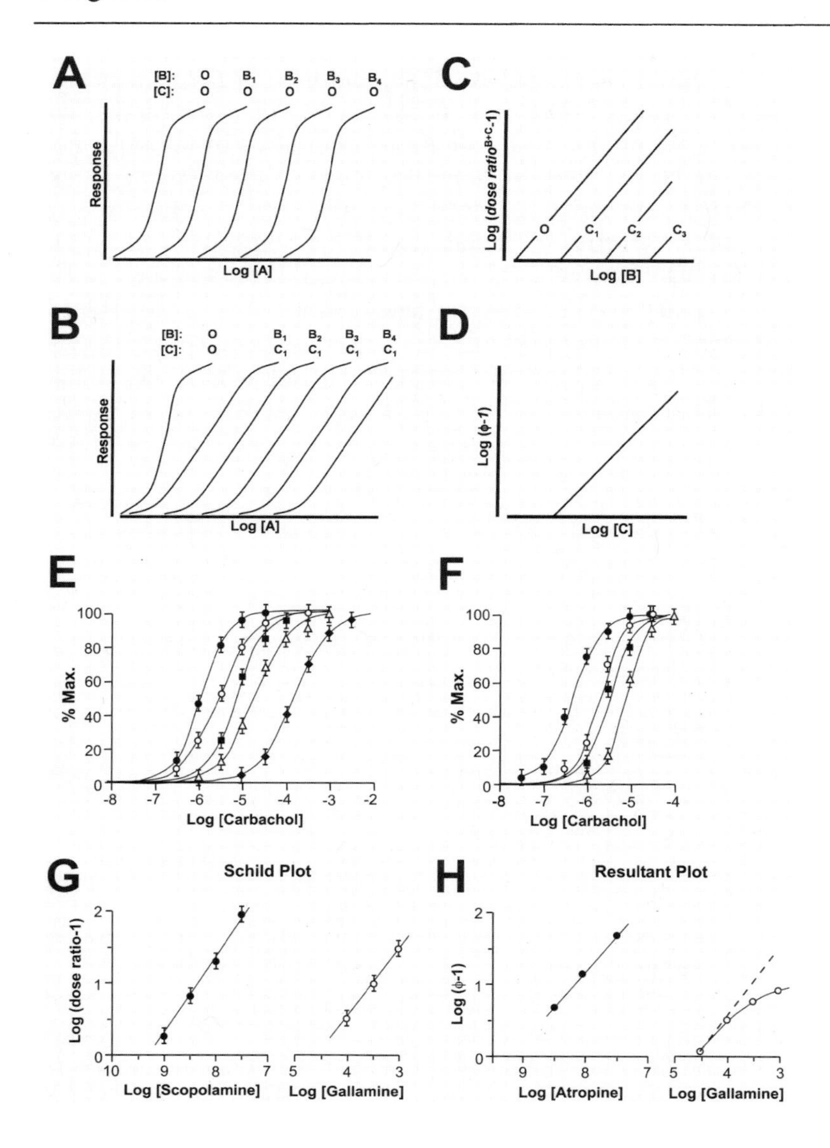

Figure 2-4. Pharmacologic resultant analysis of competitive and allosteric ligands. The derivation of ϕ and the properties of the Schild versus the resultant plot are discussed in the text. Panels A-D are modified from Black et al. (1986) and panels E-F are modified from Kenakin and Boselli (1989).

Resultant analysis is performed in the following manner (Black et al. [1986]); Kenakin [2001;2004]; and figure 2-4). The properties of a reference antagonist, denoted "*B*," known to be a competitive antagonist and presumably free from secondary, or "resultant," activity, are compared to those of a test compound, denoted as "*C*," which is suspected of giving a blocking effect resulting from competitive antagonism plus additional actions. First, the response to agonist is examined alone, and then in the presence of the test antagonist, *B* (figure 2-4A). In a second series of experiments, the agonist is examined in the presence of a single concentration of the reference antagonist (C_1) and the same increasing concentrations of the test antagonist, *B* (figure 2-4B). This second series is repeated for different concentrations of *C* (C_1, C_2, C_3) to produce a series of "Schild lines," i.e. log ($dose\ ratio^{B+C}$-1) (figure 2-4C). The distance between each displaced Schild line and the control plot (C=0) on the log [*B*] axis is measured and defined as log (ϕ). Replot values of ϕ for different concentrations of *C* are then performed (figure 2-4D). If *C* is competitive or has a competitive property among other independent properties, then $y=1+[C]/K_{D_C}$ and a plot of (ϕ-1)/log [*C*] will yield the K_D for *C* (K_{D_C}).

The derivation of the equations, and the assumptions therein, are as follows. In simple competitive antagonism:

$$\text{response}^B = f\frac{[A^B]}{K_{D_A}\left(1+\frac{B}{K_{D_B}}\right)+[A^B]} \qquad (2.22)$$

where response in the presence of a competitive (i.e. "reference") antagonist, *B*, is shown by the Gaddum equation, and $[A^B]$ signifies the concentration of agonist, *A*, evaluated in the presence of the test antagonist, *B*. By definition, the complex relationship between occupancy and response, denoted *f*, is not changed by *B*, allowing the cancellation of this function in the comparison of agonist response in the absence and presence of *B*, as described in equation 2.18 ($\text{dose ratio} = 1+\frac{[B]}{K_{D_B}}$), for the Schild equation. However, blockers of receptor response not only may be competitive antagonists at the orthosteric site but also may be allosteric in nature, and may change the relationship, *f*, between occupancy and response. Black et al. (1986) noted that when a ligand, *C*, both competes with *A* *and* alters the transducer function, *f*, then equation 2.22 is modified as follows:

$$\text{response}^{C} = f^{C}\left(\frac{\left[A^{C}\right]}{K_{D_A}\left(1+\frac{[C]}{K_{D_C}}\right)+\left[A^{C}\right]}\right) \quad (2.23)$$

When equal responses to *A* are observed in the absence and presence of *C*, f and f^{C} cannot cancel. The solution of equation 2.23, however, comes from the additive rule of Paton and Rang (1965) for two null, competitive antagonists acting at the orthosteric site for agonist:

$$\text{response}^{B+C} = f\frac{\left[A^{B+C}\right]}{K_{D_A}\left(1+\frac{[B]}{K_{D_B}}+\frac{[C]}{K_{D_C}}\right)+\left[A^{B+C}\right]} \quad (2.24)$$

Equation 2.24 represents the case when *B* and *C* are both competitive antagonists. However, if *C* produces a "resultant" effect of competitive antagonism plus other effects, then equation 2.24 is modified to:

$$\text{response}^{B+C} = f^{C}\frac{\left[A^{B+C}\right]}{K_{D_A}\left(1+\frac{[B]}{K_{D_B}}+\frac{[C]}{K_{D_C}}\right)+\left[A^{B+C}\right]} \quad (2.25)$$

The relationship between occupancy and response in the presence of *C* does not cancel if one is comparing equation 2.22 with equation 2.25; it *does* cancel when comparing 2.23 and 2.25. Consequently for comparing equal responses of agonist in the presence of *C* versus *C* + *B*, $\text{response}^{C}=\text{response}^{B+C}$ and:

$$f^C\left(\frac{[A^C]}{K_{D_A}\left(1+\frac{[C]}{K_{D_C}}\right)+[A^C]}\right)=f^C\left(\frac{[A^{B+C}]}{K_{D_A}\left(1+\frac{[B]}{K_{D_B}}+\frac{[C]}{K_{D_C}}\right)+[A^{B+C}]}\right)$$

(2.26)

and with elimination of f^C and rearrangement:

$$\text{dose ratio}^{B+C}=\frac{[A^{B+C}]}{[A^C]}=1+\frac{[B]}{K_{D_B}\left(1+\frac{[C]}{K_{D_C}}\right)} \quad (2.27)$$

where dose ratio^{B+C} is the concentration ratio of agonist, *A*, required to surmount both $B+C$ versus *C* alone

As noted above, Schild plots for the test antagonist (figure 2-4A) alone and for the test antagonist plus a range of reference antagonist (figure 2-4B) are obtained. Equieffective dose ratios are compared. A term ϕ is defined as the ratio of reference antagonist concentrations giving equal dose ratio-1 values in the presence of various concentrations of test antagonist:

$$\log(\phi-1)=\log[C]-\log K_{D_C} \quad (2.28)$$

and a replot, as in figure 2-4D, is obtained. Thus, if the test antagonist, *C*, is a simple competitive antagonist, a plot of log (ϕ-1) versus log [*C*] will yield a straight line, with a slope of 1. If the linear regressions can be fit to a common slope of 1.0 (i.e. if 1.0 is within the 95% confidence limits of each of the slopes), then a refit of the data in figure 2-4D to a slope of unity yields the K_{D_C} for the reference antagonist, *C*. Deviation from unity or a curvilinear relationship between log (ϕ-1) and log [*C*] is evidence of non-orthosteric antagonistic effects. As shown in figure 2-4G versus 2-4H, the resultant plot (2-4H) allows discrimination of gallamine as an allosteric inhibitor of carachol actions at the muscarinic receptors.

If the two agents both are null, or simple, competitive antagonists, then the reduction in response to agonist by these two antagonists is a function of a factor equal to the additive concentration of antagonists, expressed as a fraction of their K_{D_B} values:

$$\frac{[B_1]}{K_{D_{B_1}}} + \frac{[B_2]}{K_{D_{B_2}}} \text{(i.e. additive dose ratios)} \tag{2.29}$$

If the two antagonists are both null antagonists, the effect of the two antagonists on response will be additive and can be calculated by the Gaddum equation (2.16). However, if the two antagonists are not interacting at the same site on the receptor, then additive dose ratios likely will not be observed.

SUMMARY

The methods summarized in this chapter provide approaches for characterizing the specificity of a drug or hormone effect. This specificity provides evidence for a receptor-mediated response. Once a receptor is implicated in a response, a variety of analyses are available for estimating the equilibrium dissociation constants for agonists, partial agonists, antagonists and inverse agonists. It is clear that evaluation of receptor-mediated response provides a great deal of information, both qualitatively and quantitatively, regarding ligand-receptor interactions and subsequent receptor-activated responses. These approaches were developed when the biological preparation under study was intact tissue or native target cells. However, in current high throughput screening technologies where the read-out is a cell-based response, these strategies have comparable value in elucidating the mechanism and providing quantitative parameters to describe the properties of novel agents. In fact, the only receptor parameter that cannot be obtained from studies of dose-response relationships in intact cell or tissue preparations is receptor density. Radioligand binding methods for characterization of receptors yielding information concerning receptor specificity, affinity *and* density are discussed in chapter 3.

REFERENCES

General

Ahlquist, R.P. (1948) A study of the adrenotropic receptors. Am. J. Physiol. 155: 586-600.

Arunlakshana, O. and Schild, H.O. (1959) Some quantitative uses of drug antagonists. Brit. J. Pharm. 14:48-58.

Barlow, R., Scott, N.C. and Stephenson, R.P. (1967) Brit. J. Pharmacol. 31:188-196. Colquhoun, D. (1973) The relation between classical and cooperative models for drug action. In *Drug Receptors*, H.P. Rang (ed.), pp. 149-182. Baltimore: University Park Press.

Black, J.W., Gershkowitz, V.P., Leff, P. and Shankley, N.P. (1986) Analysis of competitive antagonism when this property occurs as part of a pharmacological resultant. Brit. J. Pharm. 89:547-555.

DeLean, A., Stadel, J.M. and Lefkowitz, R.J. (1980) A ternary complex model explains the agonist-specific binding properties of the adenylate cyclase-coupled β-adrenergic receptor. J. Biol. Chem. 255:7108-7117.

Furchgott, R.F. (1972) The classification of adrenoceptors (adrenergic receptors): An evaluation from the standpoint of receptor theory. In *Handbuch der Experimentellen Pharmakologie*, vol. 33, *Catecholamines*, H. Blashko and E. Muscholl (eds.), pp. 283-335. Berlin: Springer-Verlag.

Furchgott, R.F. (1978) Pharmacological characterization of receptors: Its relation to radioligand-binding studies. Fed. Proc. 37:115-120.

Gaddum, J.H. (1937) The quantitative effects of antagonist drugs. J. Physiol. (London) 89:79-9P.

Gaddum, J.H. (1943) Biological aspects: The antagonism of drugs. Trans. Faraday Soc. 39:323-333.

Gaddum, J.H. (1957) Theories of drug antagonism. Pharm. Rev. 9:211-217.

Kenakin, T.P. (1984a) The classification of drugs and drug receptors in isolated tissues. Pharmacol. Rev. 36:165-222.

Kenakin, T.P. (1987a) *The Pharmacological Analysis of Drug Receptor Interaction.* NY: Raven Press.

Kenakin, T.P. (1987b) What can we learn from models of complex drug antagonism in classifying hormone receptors? In *Perspectives on Receptor Classification, Receptor Biochemistry and Methodology*, vol. 6, pp. 167-185. Black, J.W., Jenkinson, D.H. and Gershkowitz, V.P. (eds.). NY: Alan R. Liss.

Kenakin, T.P. (2001) Quantitation in Receptor Pharmacology. Rec. & Chan. 7:371-385.

Kenakin, T.P. (2004) *A Pharmacology Primer: Theory, Methods, and Application.* San Diego: Elsevier Press, Inc.

MacKay, D. (1966) A general analysis of the receptor-drug interaction. Brit. J. Pharmacol. 26:9-16.

MacKay, D. (1978) How should values of pA_2 and affinity constants for pharmacological competitive antagonists be evaluated? J. Pharm. Pharmac. 30:312-313.

MacKay, D. (1982) Dose-response curves and mechanisms of drug action. Trends in Pharm. Sci. 2:496-499.

Parker, R.B. and Waud, D.R. (1971) Pharmacological estimation of drug-receptor dissociation constants: Statistical evaluation. I. Agonists. J. Pharm. Exp. Ther. 177:1-12.

Schild, H.O. (1947) pA, a new scale for the measurement of drug antagonism. Brit. J. Pharm. 2:189-206.

Schild, H.O. (1957) Drug antagonism and pA_x. Pharm. Rev. 9:242-245.

Schild, H.O. (1973) Receptor classification with special reference to β-adrenergic receptors. In *Drug Receptors*, pp. 29-36, H.P. Rang (ed.). Baltimore: University Park Press.

Stephenson, R.P. (1956) A modification of receptor theory. Brit. J. Pharm. 11: 379-393.

Thron, C.D. (1973) On the analysis of pharmacological experiments in terms of an allosteric receptor model. Mol. Pharmacol 9:1-9.

Van Rossum, J.M. (1963) Cumulative dose-response curves. II. Technique for the making of dose-response curves in isolated organs and the evaluation of drug parameters. Arch. Int. Pharmacodyn. Ther. 143:299-330.

Waud, D.R. (1969) On the measurement of the affinity of partial agonists for receptors. J. Pharm. Exp. Ther. 170:117-122.

Specific

Berg K.A., Maayani, S., Goldfarb, J., Scaramellini, C., Leff, P. and Clarke, W.P. (1998) Effector Pathway-Dependent Relative Efficacy at Serotonin Type 2A and 2C Receptors: Evidence for Agonist-Directed Trafficking of Receptor Stimulus. Mol. Pharm. 54:94-104.

Berthelson, S. and Pettinger, W.A. (1977) A functional basis for classification of α-adrenergic receptors. Life Sci. 21:595-606.

del Castillo, J. and Katz, B. (1957) Interaction at endplate receptors between different choline derivatives. Proc. Royal Soc. Lond., Series B, 146:369-381.

Furchgott, R.F. (1954) Dibenamine blockade in strips of rabbit aorta and its use in differentiating receptors. J. Pharm. Exp. Ther. 111:265-284.

Furchgott, R.F. (1966) The use of β-haloalkylamines in the differentiation of receptors and in the determination of dissociation constants of receptor-agonist complexes. In *Advances in Drug Research*, vol. 3, pp. 21-55. Harper, N.J. and Simmonds, A.B. (eds.). New York: Academic Press.

Furchgott, R.F. and Bursztyn, P. (1967) Comparison of dissociation constants and relative efficacies of selective agonists acting on parasympathetic receptors. Ann. NY Acad. Sci. 139:882-889.

Kaumann, A.J. and Marano, M. (1982) On equilibrium dissociation constants for complexes of drug receptor subtypes: selective and nonselective interactions of partial agonists with two β-adrenoceptor subtypes mediating positive chronotropic effects of (-) isoprenaline in kitten atria. Naunyn Schmiedebergs Arch. Pharmacol. 219:216-221.

Kenakin, T.P. (1984b) The relative contribution of affinity and efficacy to agonist activity: organ selectivity of noradrenaline and oxymetazoline with reference to the classification of drug receptors. Br. J. Pharmacol. 81(1):131-141.

Kenakin, T.P. and Boselli, C. (1989) Pharmacologic discrimination between receptor heterogeneity and allosteric interaction: resultant analysis of gallamine and pirenzepine antagonism of muscarinic responses in rat trachea. J. Pharmacol. Exp. Ther. 250(3):944-952.

Kono, T. and Barham, F.W. (1971) The relationship between the insulin-binding capacity of fat cells and the cellular response to insulin. J. Biol. Chem. 246: 6210-6216.

Kuwasako, K., Cao Y-N., Nagoshi, Y., Tsuruda, T., Kitamura, K. and Eto, T. (2004) Characterization of the Human Calcitonin Gene-Related Peptide Receptor Subtypes Associated with Receptor Activity-Modifying Proteins. Mol. Pharm. 65:207-213.

Langer, S.Z. and Trendelenberg, U. (1960) The effect of a saturable uptake mechanism on the slope of dose-response curves for sympathomimetic amines and on the shifts of dose-response curves produced by a competitive antagonist. J. Pharm. Exp. Ther. 167:117-142.

Leff, P., Dougall, I.G., Harper, D.H. and Dainty, I.A. (1990) Errors in agonist affinity estimation: Do they and should they occur in isolated tissue experiments? Trends in Pharm. Sci. 11:64-67.

Lefkowitz, R.J., Cotecchia, S., Samama, P. and Costa, T. (1993) Constitutive activity of receptors coupled to guanine nucleotide regulatory proteins. Trends in Pharm. Sci. 14:303-307.

Lin, C.W. and Musacchio, J.M. (1983) The determination of dissociation constants for substance P and substance P analogues in the guinea pig ileum by pharmacological procedures. Mol. Pharmacol. 23:558-562.

Paton, W.D.M. and Rang, H.P (1965) The uptake of atropine and related drugs by intestinal smooth muscle of the guinea pig in relation to acetylcholine receptors. Proc. R. Soc. Lond. B. 163:1-44.

Schultz, R., Wuster, M., Krenss, H. and Hers, A. (1980) Lack of cross-tolerance on multiple opiate receptors in the vas deferens. Mol. Pharmacol. 18:395-401.

Schütz, W. and Freissmuth, W.B. (1992) Reverse intrinsic activity of antagonists on G protein-coupled receptors. Trends in Pharm. Sci. 13:376-380.

Sine, S.M. and Taylor, P. (1981) Relationship between reversible antagonist occupancy and the functional capacity of the acetylcholine receptor. J. Biol. Chem. 256:6692-6699.

Tallarida, R.J., Cowan, A. and Adler, M.W. (1979) pA_2 and receptor differentiation: A statistical analysis of competitive antagonism (minireview). Life Sci. 25:637-654.

3. IDENTIFICATION OF RECEPTORS USING DIRECT RADIOLIGAND BINDING TECHNIQUES

The availability of radioactively labeled hormones and drugs that retain their biological activity has allowed the direct identification of binding sites for these agents in target tissues and cells where they elicit a response. This chapter will describe methods for the identification of radioligand binding sites and for obtaining data to establish whether the observed binding sites represent the physiological "receptor" for a particular hormone, neurotransmitter, or drug. The same conceptual approach, however, also can be applied to studies of receptor identification when the ligand being monitored is fluorescent or otherwise detectable, rather than radioactive.

METHODS--DATA GENERATION

The interaction one wishes to study is the binding of drug or hormone, D, to its receptor, R, to form a drug-receptor complex, DR. A reaction involving the binding of a single ligand to a single population of homogenous binding sites by a fully reversible reaction is said to behave via the principles of mass action when the rates of the reaction are driven by the quantities (mass) of reactants in each side of the equilibrium, i.e., $[D]$, $[R]$ and $[DR]$, as described in equation 3.1:

$$D + R \underset{k_2}{\overset{k_1}{\rightleftharpoons}} DR \tag{3.1}$$

where

D	=	hormone or drug;
later, $^*D$	=	radiolabeled hormone or durg
R	=	receptor
DR	=	complex of receptor with drug
k_1	=	association rate constant
k_2	=	dissociation rate constant
$k_1[D][R]$	=	initial forward or association rate (3.2)
$k_2[DR]$	=	initial reverse or dissociation rate (when studied under conditions where no detectable association occurs, as discussed later) (3.3)

The following discussion will consider the choice of radioligand $^*D$, the biological source of R, and variables in the incubation that will optimize detection of the radiolabeled drug-receptor complex, $^*DR$.

Choice of a Radioligand

One of the most critical considerations in the successful application of radioligand binding techniques is the choice of an appropriate isotope for the radiolabeling of the ligand, D. Table 3-1 outlines the characteristics of the radioisotopes commonly employed for receptor identification. These characteristics ultimately affect the choice one makes for radiolabeling the ligand. Tritium is a popular choice because its long half-life means that the ligand does not have to be resynthesized or repurchased frequently, although it may need to be repurified frequently to remove tritium that has exchanged with water or to remove radiation-induced inactivation products of the ligand. However, a simple calculation may emphasize why tritiated ligands are only suitable when there is an ample and economical supply of biological material that possesses the putative receptor:

$[R]$ is often only 0.1-0.2 pmol/mg protein in crude homogenates or 1 pmol/mg protein in purified plasma membranes for cell surface receptors expressed at 1 R/μm^2 on the cell surface, a frequently encountered receptor density.

$^3H \cdot D$ at a specific radioactivity of 29 Ci/mmol (cf. table 3-1) counted at 45% counter efficiency means that there are 29 cpm/fmol of radioligand (1 μCi = 2.2×10^6 dpm; dpm × efficiency [0.45] = cpm)

When [R] = 0.1 pmol/mg, then at *maximal* receptor occupancy (which one never achieves experimentally) the binding detected would be 2,900 cpm/incubation containing 1 mg of membrane protein. Receptor densities in native tissues often exist at ~0.1 pmol/mg, or even less, such that receptor quantity *is* limiting. With the availability of cDNAs encoding most receptors of interest to investigators, the limitation of receptor availability is no longer an issue, if the characterization is directed at the receptor *per se*, rather than at a target tissue of particular biological interest.

Iodinated ligands, because of their higher specific radioactivity (i.e., 2,200 Ci/mmol), are especially valuable when identifying receptors where only trace biological material is available, e.g in isolated nuclei of the central nervous system.

Table 3-1. Characteristics of radioisotopes that influence their utility for identification of receptors.

Isotope	*Specific Radioactivity*[1] *Ci/mmol*	*Half-life*	*Other Considerations*
^3H	29.4	12.3 years	Bioactivity usually unchanged by tritiation; can introduce >1 mole ^3H/mole ligand to increase specific radioactivity
^{125}I	2125	60.2 days	Require tyrosine or unsaturated cyclic system in ligand structure to achieve incorporation of ^{125}I, except in unusual circumstances (Bearer et al. [1980]); high specific radioactivity especially useful when receptor availability limited
^{32}P	9760	14.2 days	Short half-life is a technical frustration
^{35}S	4200	86.7 days	Good sensitivity; ^{35}S must be added during appropriate step in chemical synthesis
^{14}C	0.064	5568. years	Exceedingly poor sensitivity because of low specific radioactivity

[1]Specific radioactivity indicated assumes incorporation of one mole of radioisotope into one mole of ligand.

Note: Values for specific radioactivity and half-life taken from Kobayashi, Y. and Maudsley, D.V. (1974), *Biological Applications of Liquid Scintillation Counting*. New York: Academic Press.

The Incubation

The obvious additions to the binding incubation include a suitable preparation of receptor-containing biological material (e.g., intact cells, homogenates, isolated membrane fractions) and the radioligand chosen to identify the receptor of interest. The concentration of materials added and the duration of the incubation essential for obtaining detectable binding are influenced by some of the consequences of mass action:

At equilibrium, the rate of association = rate of dissociation
$k_1[D][R] = k_2[DR]$ (3.4)
and the ratio of bound D (as DR complex) to the reactants is:

$$\frac{[DR]}{[D][R]} = \frac{k_1}{k_2} = K_A, \text{molar}^{-1} \quad (3.5)$$

K_A = EQUILIBRIUM ASSOCIATION CONSTANT

However, K_D, the EQUILIBRIUM DISSOCIATION CONSTANT, is used most commonly to describe the affinity of receptor R for drug D because the units are molar, rather than reciprocal molar. The K_D represents the concentration of a drug or hormone that half-maximally occupies the receptor at equilibrium (see later discussion related to figure 3.3):

$$K_D, \text{molar} = \frac{[D][R]}{[DR]} = \frac{k_2}{k_1} \quad (3.6)$$

It should be emphasized that an *increase* in K_D correlates with a *decrease* in receptor affinity.

The incubation is continued until sufficient $^*DR$ has accumulated so that the quantity of bound radioligand is detectable. The overall rate of association can be increased by increasing the concentration of either or both of the reactants, $^*D$ and R. These manipulations also will increase the quantity of $^*DR$ formed at equilibrium, and may lengthen the time to reach equilibrium. Optimal incubation conditions are determined empirically for each system and are chosen to maximize formation of $^*DR$ while minimizing the degradation of $^*D$ and R.

Separation of Bound from Free Radioligand

To determine the quantity of $^*DR$ accumulated, one must have a method for terminating the incubation that permits the resolution of $^*D$ from $^*DR$. There are several methods available for the resolution of $^*D$ and $^*DR$. Equilibrium dialysis is considered the *theoretically* most attractive approach, because assessing bound versus free ligand does not require perturbing the equilibrium. In contrast, centrifugation and vacuum filtration, the latter being the most common method to terminate radioligand binding assays, do perturb equilibrium, since resolving $^*DR$ *from* $^*D$ results in an opportunity for $^*DR$ to dissociate, thus underestimating the amount of radioligand bound during the incubation. (Methods to separate receptor-bound from free radioligand in detergent extracts, for the study of isolated receptors, are described in chapter 5). The main parameters that determine whether or not the $^*DR$ complex will dissociate during the process of separation of $^*DR$ from $^*D$ is the time required for the separation to be completed and the affinity of the receptor for the radioligand. Remember from equation 3.6 that the K_D is a ratio of k_2/k_1. If it is assumed that the rate constant for association of the radioligand with the receptor is 10^6 M^{-1} sec^{-1}, a commonly determined value for the binding of small ligands to proteins, and that the dissociation rate is monoexponential as expected for dissociation of $^*DR$ to $^*D + R$, then

$$K_D = \frac{[*D][R]}{[*D]} = \frac{k_2}{k_1} = \frac{0.693/t_{1/2}}{10^6 \mathrm{M}^{-1}\,\mathrm{sec}^{-1}} = \frac{6.93 \times 10^{-7}\,\mathrm{M}\cdot\mathrm{sec}}{t_{1/2}} \qquad (3.7)$$

It can be calculated that a separation procedure that is complete within 0.15 $t_{1/2}$ must be utilized to avoid losing more than 10% of the $^*DR$ complex. The relationship between K_D values and allowable separation time can be calculated to avoid a loss of more than 10% of [$^*DR$], and is shown in table 3-2 (from Yamamura et al. [1985]). If the K_D for ligand binding to the receptor is 10^{-8} M, then the separation must be complete within 10 seconds. If, for example, a centrifugation assay requires one minute to thoroughly pellet the particulate material, then the receptor must interact with the radioligand with an affinity in the nanomolar range. Typically, centrifugation using "microfuges" can pellet the bulk of the particulate material within 5 seconds, but this rate of membrane pelleting must be determined empirically by the investigator for each particulate preparation. It must be remembered, however, that the above calculations are based on the assumption that the association rate constant is 10^6 M^{-1} sec^{-1}. If it is determined that the radioligand associates with the receptor with a slower rate constant, then the dissociation rate constant will be proportionately slower to obtain the same K_D, thus allowing complexes of lower affinity to be trapped without significant dissociation during the separation procedure. In addition, if the incubation is cooled to 4°C just prior to centrifugation (for example, by dilution with ice-cold buffer), then the rate of radioligand dissociation typically is slowed considerably as a result of simple kinetic molecular theory. However, this latter manipulation may introduce additional complications if the K_D value for receptor-ligand interactions is significantly different at 4°C than at the temperature at which incubation was performed.

Equilibrium Dialysis

As indicated above, equilibrium dialysis is theoretically the most accurate way to determine $^*D$ and $^*DR$, because equilibrium is not disturbed when samples are taken. A schematic diagram of an equilibrium dialysis cell is provided in figure 3-1.

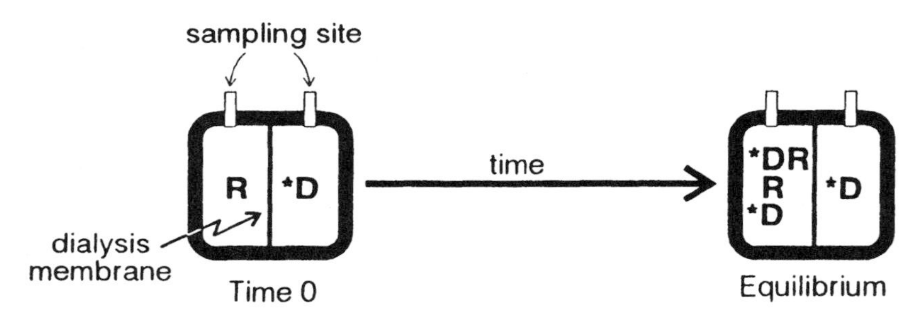

At equilibrium, the [*D] on each side of the dialysis membrane is equal

	Membrane (R) side:	[*DR] + [*D]
−	Non-membrane side:	[*D]
=	difference	= [*DR], the concentration of bound radioligand

Figure 3-1. A schematic representation of equilibrium dialysis. $[^*D]$ = concentration of free radioligand, R = receptor, and $[^*DR]$ = concentration of radioligand-receptor complex.

Dialysis cells contain two independent chambers of a known volume (e.g., 1 ml) separated by a dialysis membrane across which the radioligand $^*D$ can readily diffuse but the receptor R cannot. Radioligand is added either to one or to both chambers. The incubation is continued until "equilibrium" is attained, i.e., a longer incubation does not change the distribution of the radioligand. It is worth mentioning at this point that for most studies of radioligand binding, the term **equilibrium** is an overstatement about what is known. Equilibrium is felt to prevail when incubations of longer duration do not result in further accumulation of $^*DR$. However, investigators rarely simultaneously monitor the degradation of $^*D$ and R. Consequently, it is more appropriate to refer to attaining "**steady state**" than to attaining equilibrium, unless appropriate documentation of the lack of degradation of $^*D$ and R is provided. Since, at steady state, the concentration of free radioligand is equal on both sides of the dialysis membrane, the radioactivity detected on the side containing receptor is due to the sum of $[^*DR] + [^*D]$ and the radioactivity detected on the side of the dialysis cell containing no receptor is attributable solely to $[^*D]$. Therefore, the difference in cpm detected in the two chambers provides an estimate of the concentration of the radioligand-receptor complex $[^*DR]$.

The advantages to equilibrium dialysis are mostly theoretical, i.e. not perturbing equilibrium when an aliquot of the incubation is removed simultaneously from each chamber to assess binding at particular time points, equilibrium is not disturbed. Because the equations used to obtain binding parameters such as K_D values and total receptor densities are based on the assumption that equilibrium prevails, data obtained from equilibrium dialysis studies can be analyzed using these transformations without concern for artifacts introduced by perturbation of equilibrium that occurs upon separation of bound and free radioligand using other strategies. Theoretically, an additional advantage to equilibrium dialysis is that this strategy should be useful when the binding site for the radioligand possesses a relatively low affinity. In this situation, the radioligand likely would dissociate from the $^*DR$ complex during the time required to separate receptor-bound from free radioligand using vacuum filtration or centrifugation. However, even using equilibrium dialysis, it can be calculated that a high concentration of nearly purified receptor would be necessary in order to detect binding to receptors possessing affinities in the micromolar range.

The disadvantages of equilibrium dialysis are several. The most significant detraction to the method of equilibrium dialysis is that the cpm of $[^*DR]$ detected are typically a very small "signal" above a very high "background," which are the cpm due to the quantity of $[^*D]$ on both sides of the dialysis membrane. Consequently, obtaining statistically significant data requires many independent experiments. Thus, although equilibrium dialysis has been used as the "gold standard" for assessing the validity of other less cumbersome methods to separate free from bound radioligand, equilibrium dialysis rarely is used for the routine identification of receptors in biological membranes.

The equilibrium dialysis technique can be modified to provide determinations of ligand binding over a more rapid time frame (Colowick and

Womack [1969]). In this approach, a dialysis cell is devised so that the chamber containing the biological preparation is separated by a dialysis membrane from a buffer solution which is flowing at a constant rate. When the ligand is added to the receptor-containing preparation, it passes through to the buffer chamber at a rate that is proportional to the concentration of free ligand in the receptor-containing side of the chamber. Calibration of the ligand diffusion rate in the absence and presence of the receptor preparation allows one to obtain values for free and bound ligand. Remy and Buc (1970) have shown that steady state may be reached within less than a minute with small ligands and specially prepared dialysis membranes. Furthermore, the Colowick and Womack procedure allows utilization of the same biological preparation to determine binding at several ligand concentrations, rather than committing a large quantity of biological material to the determination of a single data point.

Centrifugation

In the centrifugation technique, the $[^*DR]$ formed during the incubation is determined by pelleting the membranes, leaving $^*D$ in the supernatant. When experimental conditions are developed that allow separation of bound from free radioligand without dissociation of the $^*DR$ complex, the cpm of $^*D$ associated with the pellet are taken to be representative of bound radioligand (less trapped $^*D$) and the cpm in the supernatant are a measure of free radioligand. Disposable microfuge tubes often are employed for these studies so that the cpm associated with the pellet can be determined simply by cutting off the tip of the microfuge tube and determining the radioactivity associated with the pellet.

One of two possible methods for centrifugation typically is employed for separating bound from free radioligand (figure 3-2). In the simplest technical approach, an aliquot of the incubation is transferred to a centrifugation tube and the particulate material is pelleted. Although this method does not seriously affect equilibrium during the pelleting ($[R]$ is changing, but $[^*D]$ is not), it does result in high background radioactivity due to the trapping of radioligand in the pellet. Washing the pellet to remove trapped $^*D$ may seriously affect the $[^*DR]$ detected, since this complex will have the opportunity to dissociate during the washing. Pellet washing consequently is an unsuitable way to decrease background radioactivity due to trapped radioligand. The alternative centrifugation approach involves layering the contents of the incubation above a solution of high density, e.g., sucrose or oil, in the centrifuge tube. The density of the solution is chosen so that it

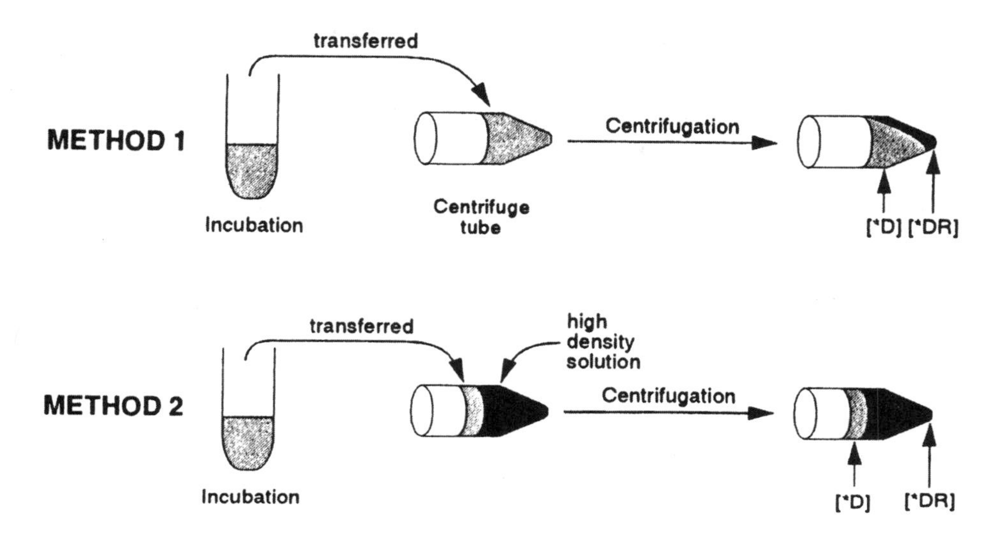

Figure 3-2. Centrifugation as a means to terminate radioligand binding assays and determine the quantity of $^*DR$ formed.

allows the immediate sedimentation of the particulate receptor-containing preparation but retains unbound $^*D$ above the higher density material in the centrifuge tube. This separation is better achieved if the radioligand is not readily miscible into the material chosen for preparing the high-density solution. Centrifuging through a high-density solution reduces background radioactivity because less radioligand is trapped in the pellet. However, a potential disadvantage of this method is that there is a greater opportunity for $[^*DR]$ to dissociate during the centrifugation process, because no radioligand is added to the intervening medium of high density, and thus both $[^*D]$ and $[R]$ are changing during the sedimentation.

Vacuum Filtration

Vacuum filtration is the most common method for separating receptor-bound from free radioligand, primarily because of the relative ease of handling a large number of samples as well as the commercial availability of a variety of filtration devices. When an incubation is terminated by pouring the incubate through a filter under vacuum, the membranes are retained by the filter, and the free radioligand passes through the filter. The filter composition is chosen so that it retains all of the membranes and binds very little, if any, free radioligand. Filters can be coated to prevent or reduce radioligand adsorption, e.g., with bovine serum albumin or gelatin when peptides are the ligands, or with siliconizing or other coating solutions when nonspecific binding properties of the filter are suspected to account for the adsorption of radioligand. When the affinity of the receptor for the radioligand is high enough, the filters can be washed with buffer several times to decrease background binding due to trapped radioligand. Filtration usually takes approximately 2-5 seconds/buffer wash, which means that the minimum affinity (maximum K_D) of a receptor for a ligand that can be identified using vacuum filtration techniques is 10^{-8} M (cf. table 3-2).

Table 3-2. Relationship between equilibrium dissociation constant (K_D) and allowable separation time of *DR* complex to avoid loss of more than 10% of $^{*}DR$.

$K_D(M)$	*Allowable Separation Time* $(0.15t_{1/2})$[1]
10^{-12}	1.2 days
10^{-11}	2.9 hr
10^{-10}	17.0 min
10^{-9}	1.7 min
10^{-8}	10.0 sec
10^{-7}	1.0 sec
10^{-6}	0.1 sec

[1]Calculations of $t_{1/2}$ (half-life for dissociation) assuming an association rate constant of 10^{6} M^{-1} sec^{-1} as shown in equation 3.7.

Note: This table was revised from Yamamura, H.I., Enna, S.J. and Kuhar, M.J. (1985) *Neurotransmitter Receptor Binding*, 2nd ed. NY:Raven Press, and corrected for allowable separation times for K_D values at 10^{-6} and 10^{-7} M.

In a manner analogous to the discussion above for centrifugation techniques, this "cutoff" occasionally can be "stretched" by terminating the incubation by dilution with ice-cold buffer immediately prior to pouring the incubation contents over the filter which, because it is under vacuum, is also at a temperature lower than ambient temperature.

CRITERIA EXPECTED FOR BINDING OF $^{*}D$ TO THE PHYSIOLOGICALLY RELEVANT RECEPTOR, *R*

As stated at the outset of this chapter, once an investigator has obtained a radioligand and a suitable biological preparation containing the putative receptor of interest and has developed a binding assay that permits detection of the accumulation of $^{*}DR$ in a time- and concentration-dependent manner, the investigator then must provide convincing evidence that the binding site identified is indeed the physiologically relevant receptor. Below are listed the minimal criteria expected for binding to the genuine receptor.

1. The binding should be **saturable**, since a *finite* number of receptors are expected in a biological preparation.
2. The **specificity** of agents in competing with the radioligand $^{*}D$ for binding to R should parallel the specificity of hormones, drugs, or their analogs in eliciting their physiological effect via the putative receptor of interest.
3. The **kinetics** of binding should be **consistent** with the time-course of the biological effect elicited by D, and there should be internal consistency between the K_D value for radioligand determined using steady state incubations and the value calculated from the ratio of rate constants (k_2/k_1) determined in kinetic experiments.

Knowing other salient features that characterize the ligand-induced biological effect may allow additional criteria to be established that permit further documentation of the reliability of the radioligand binding data. For example, adrenergic agents elicit their effects stereoselectively, i.e., (-) or (*l*) isomers of catecholamines are considerably more potent than their corresponding (+) or (*d*) isomers. Consequently, one expects that the binding of radiolabeled agents to adrenergic receptors and competition for that binding similarly will be stereoselective in nature.

Determining the Saturability of Radioligand Binding

To determine whether or not binding is saturable, incubations are performed with increasing concentrations of radioligand to determine whether binding eventually "plateaus." Before describing the data obtained and the graphical transformations used to analyze the data, it is helpful to consider again consequences of a fully reversible bimolecular reaction driven by mass action.

$$D + R \underset{k_2}{\overset{k_1}{\rightleftharpoons}} DR$$

Definitions: $[D]$ = free ligand concentration, later denoted as F

$[R]$ = concentration of unoccupied receptors

$[DR]$ = concentration of ligand-receptor complex. Since this-by definition-is equal to the concentration of ligand **bound** to R, $[DR]$ is later denoted as B

$[R]_{TOT}$ = total concentration of receptor sites = $B_{\max}$ = $[R]$ + $[DR]$

The principles of the model of mass action are: 1) all receptors are equally accessible to ligand; 2) all receptors are either free or bound (the model ignores states of partial binding); 3) neither ligand nor receptor is altered by binding; 4) binding is fully reversible.

As shown earlier (equations 3.5 and 3.6),

$$K_D = \text{equilibrium dissociation constant} = \frac{[D][R]}{[DR]} = \frac{k_2}{k_1} = \text{molar units}$$

$$K_A = \text{equilibrium association constant} = \frac{[D][R]}{[D][R]} = \frac{k_1}{k_2} = \text{units of molar}^{-1}$$

Y = fractional saturation of R with D

$$Y = \frac{[DR]}{[R]_{TOT}} = \frac{B}{B_{\max}} \tag{3.8}$$

and

$$B = [DR] = Y \cdot [R]_{TOT}$$

$$(1 - Y) = \text{fraction of unoccupied receptors} \tag{3.9}$$

and

$$[R] = (1 - Y) \cdot [R]_{TOT}$$

Often it is useful to express radioligand binding data in terms of fractional occupancy, Y. To determine the relationship between Y, $[D]$, and K_D at equilibrium, the following algebraic manipulations can be performed:

$$\text{Rate of association} = k_1[D][R] = k_1 [D](1 - Y)[R]_{TOT} \tag{3.10}$$

$$\text{Rate of dissociation} = k_2[DR] = k_2 (Y)[R]_{TOT} \tag{3.11}$$

At equilibrium, the rate of association = the rate of dissociation, and:

$$k_1[D](1 - Y)[R]_{TOT} = k_2 (Y)[R]_{TOT}$$

divide through by $[R]_{TOT}$:

$k_1[D] - (k_1 [D]Y) = k_2Y$

isolate Y: $k_1[D] = Y(k_2 + k_1[D])$

$$Y = \frac{k_1[D]}{k_2 + k_1[D]}$$

divide right-hand side of equation by k_1:

$$Y = \frac{[D]}{k_2 / k_1 + [D]}$$

since

$$\frac{k_2}{k_1} = K_D$$

$$Y = \frac{[D]}{K_D + [D]} \tag{3.12}$$

To define Y in terms of K_A ($= 1/K_D$)

$$Y = \frac{K_A[D]}{1 + K_A[D]} \tag{3.13}$$

The forms of the equation

$$Y = \frac{[D]}{K_D + [D]} \text{ and } Y = \frac{K_A[D]}{1 + K_A[D]} \tag{3.14}$$

are reminiscent of the Michaelis-Menten equation when k_3, the rate of product formation, is assumed to be zero (Koshland [1970]). These equations also are formally equivalent to the adsorption isotherm derived by Langmuir (1918) for adsorption of gases to a surface at various temperatures and to the equations of A. J. Clark described in chapter 1 (see also Whitehead [1970] and Wieland and Molinoff [1987]).

To assess saturability of radioligand binding, the characteristics of binding as a function of increasing concentrations of radioligand are determined. Figure 3-3 is a schematic diagram of saturable binding plotted as $[^*DR]$ versus $[D]$. If the bound radioactivity represents binding of a single radioligand to a saturable receptor population possessing a single affinity, K_D, for the ligand, $^*D$, then the plot of $[^*DR]$ versus $[D]$ will yield a rectangular hyperbola:

Since

$$Y=\frac{[DR]}{[R]_{TOT}}=\frac{[D]}{K_D+[D]} \tag{3.15}$$

then

$$[DR]=\frac{[D][R]_{TOT}}{K_D+[D]}$$

and, for a rectangular hyperbola:

$$y=\frac{ax}{b+x}$$

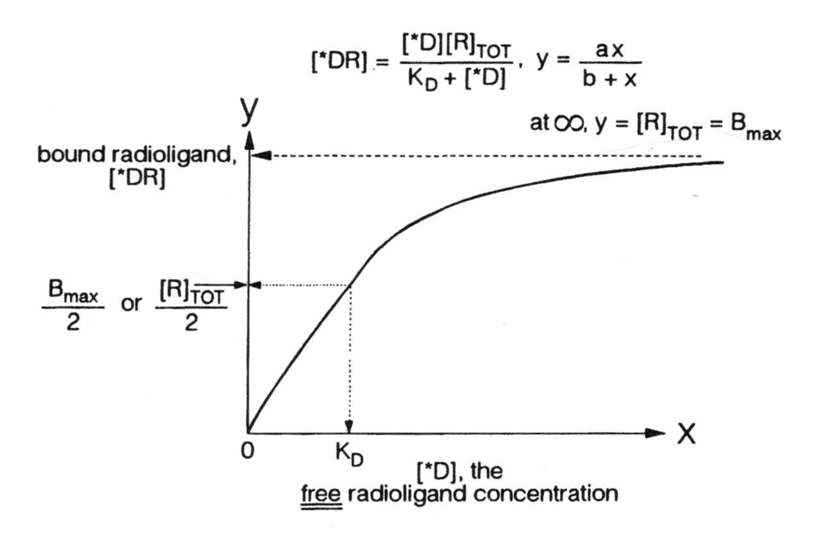

Figure 3-3. The hyperbolic relationship of mass action law: $*D+R \rightleftharpoons *DR$.

It was said earlier that K_D, a measure of the affinity of the receptor for ligand, is the concentration of ligand that half-maximally occupies the receptor. This relationship can be demonstrated by simple algebra, as shown below:

$$[DR] = \frac{[R]_{TOT}}{2}$$

Then, from equation 3.15:

$$[R]_{TOT} = \frac{2[D][R]_{TOT}}{K_D + [D]}$$

dividing through by $[R]_{TOT}$:

$$1 = \frac{2[D]}{K_D + [D]}$$

therefore

$K_D + [D] = 2[D]$ and $K_D = [D]$

Thus, $K_D = [D]$, leading to a $[DR] = \frac{[R]_{TOT}}{2}$

An important consequence of the fact that a saturation isotherm plotted as $[^*DR]$ versus $[^*D]$ is a rectangular hyperbola is that the horizontal asymptote is $[R]_{TOT}$ or B_{max}.

Since

$$[DR] = \frac{[D][R]_{TOT}}{K_D + [D]}$$

$[DR](K_D + [D]) = [D][R]_{TOT}$

$[DR]K_D + [D][DR] = [D][R]_{TOT}$

$[D]([R]_{TOT} - [DR]) = [DR]K_D$

$$[D]=\frac{[DR]K_D}{[R]_{TOT}-[DR]}$$

As $[DR] \rightarrow [R]_{TOT}$, $[D] \rightarrow \infty$ since $([R]_{TOT} - [DR]) \rightarrow 0$, and dividing a fraction by zero yields infinity.

This relationship means that B_{max} will be attained only at *infinite* concentrations of $^*D$. The importance in understanding this fundamental principle that B_{max} is a value attained only at *infinite* $[^*D]$ will become more clear when data transformations such as the Scatchard plot or Hill plot, or non-linear regression analyses of the data are dicussed later. But the simple conclusion is that an investigator will never observe B_{max} experimentally; B_{max} may be *approached* but never attained.

Defining Non-Specific Binding

Before actually plotting $[^*DR]$ versus $[^*D]$, the investigator must appreciate that all of the radioligand binding detected is not necessarily so-called *specific* binding, i.e., binding resulting from interaction of the radioligand with the physiologically relevant receptor. Some radioactivity which is detected as "bound ligand" typically represents nonreceptor binding to other sites in the membrane or trapped in pelleted material or adsorbed to filters, etc. This nonreceptor binding is referred to as "nonspecific binding." One of the most difficult challenges in developing a reliable radioligand binding assay is to determine how one will validly define nonspecific binding.

The *definition* of nonspecific binding always will be an arbitrary one. However, the validity of the definition can be strengthened if it is determined that nonspecific binding, as defined, increases linearly as a function of increasing concentrations of $^*D$. This is expected, because binding due to ligand adsorption to filters or to other similar nonreceptor sites would *not* be expected to be saturable. Another observation that bolsters confidence in the definition employed for nonspecific binding is that the *specific* binding component obtained using this definition meets all of the criteria expected for binding to the physiologically relevant receptor (specific binding = total binding - nonspecific binding).

Several approaches are used at the outset to estimate the amount of nonspecific binding that is contributing to the total amount of radioligand bound. Usually, these initial definitions are modified as more experience is obtained in studying the biological preparation to further improve the validity of the definition for nonspecific binding. Commonly employed definitions for nonspecific binding are outlined below.

1. Nonspecific binding often is defined as that binding which cannot be competed for by unlabeled ligand present at 100 × K_D for the unlabeled agent. The rationale for this choice is that if the competitor chosen for the definition of nonspecific binding is interacting with R in a bimolecular reaction driven by mass action law, R should be nearly saturated by the competitor at concentrations 10 × K_D. Thus, 100 × K_D provides a substantial excess of unlabeled competitor. However, an important and often overlooked assumption in this definition is that radioligand is present at concentrations significantly less than the K_D for radioligand and that the competitor interacts with the receptor via a bimolecular reaction obeying mass action law, such that trivial detectable radioligand binding would be expected in the presence of a competitor present at a concentration 100 × K_D for the competitor. Another limitation to this definition is that, simply by isotopic dilution, detectable radioligand binding will be decreased significantly by adding a 100-fold excess of unlabeled ligand of the same chemical structure as the radioligand. Use of a competing agent structurally distinct from that of the radioligand is preferable for defining non-specific binding.
2. An alternate method for approximating nonspecific binding is to add increasing concentrations of a competitor, X, to an incubation of membranes with a single concentration of $^*D$. When $[^*DR]$ is plotted versus $\text{Log}_{10}[X]$, the decrease in binding of $^*D$ as a function of increasing $[X]$ will appear to plateau at some concentration of X. The concentration of X where the curve becomes asymptotic with a horizontal line is taken as a concentration of X appropriate to add to incubation tubes for determination of nonspecific binding. This approach is most valid when investigators choose a competitor (such as X) that is structurally different from D but still is known to interact with the receptor of interest. This assures that the decrease in binding of $^*D$ due to the presence of unlabeled competitor is not simply a result of isotopic dilution of $^*D$ with D. Furthermore, this approach does not assume that $^*D$ or the competitor will occupy the sites as a function of mass action law, i.e., this approach allows for the possibility that negative cooperativity among sites binding $^*D$ or multiple populations of sites binding $^*D$ may exist.
3. A third approach for defining nonspecific binding is mathematical rather than experimental in nature, and can be achieved using computer-assisted analysis of radioligand binding. The mathematical model on which the computer program is based would then include, in addition to an algebraic description of radioligand interacting with *saturable* binding sites, a component of binding that is nonsaturable in nature and increases linearly with increasing concentrations of radioligand. In many ways, mathematically derived definitions of nonspecific binding are preferable because no biases exist in the definition, such as choice of the non-

radiolabeled competitor or of its concentration to be added to the incubation. Furthermore, the statistical error that occurs due to scatter in the nonspecific binding is not introduced into estimates for specific binding. However, the *mathematical* model for defining non-specific binding on which computer programs are based may not necessarily represent the *molecular* model that accurately reflects the nature of the radioligand binding to a particular biological preparation. For example, nonspecific binding is a catch-all phrase that experimentalists use to refer to any binding that is not identifying the physiologically relevant receptor. However, there are situations in which nonspecific binding, according to this definition, would include a saturable component of binding, albeit of probable lower affinity than binding to the cell surface receptor of particular interest. For example, catecholamines would be expected to interact not only with cell surface adrenergic receptors but also with intracellular enzymes that participate in catecholamine metabolism and degradation, such as catechol-o-methyl transferase. Although contributions to radiolabeled catecholamine binding by these enzymes can be minimized by the inclusion of enzyme inhibitors in the incubation, this example demonstrates that a computer program that isolates out nonsaturable binding may not necessarily define nonreceptor-related binding for the investigator.

Since it can be seen that no definition of nonspecific binding is a fail-safe means to quantify the component of total binding that does not reflect physiologically relevant receptor binding, the investigator probably will be aided best in the definition of nonspecific binding by objectively scrutinizing both empirical and mathematical approaches, and refining the definition of nonspecific binding as more insights into the biological system are obtained.

Conditions that Must Be Met to Permit Valid Interpretation of Saturation Binding Data

The development of the equations describing ligand binding to receptors, culminating in equation 3.14, involved multiple assumptions. If these assumptions are not similarly met in obtaining the raw data, then the interpretation of the data is compromised. As a reminder, these assumptions include:

1. The binding has attained equilibrium. Steady state, i.e. no net change in binding as a function of time, is not necessarily a suitable approximation for equilibrium, since steady state may mask, for example, an increase in

binding as a function of time compensating for a decrease in ligand or receptor available, due to their time-dependent degradation.

2. Bound and free radioligand can be determined accurately, and the relationship between these two is not perturbed by separation of bound from free.
3. $[R] << K_D$ for D, such that the quantity of added radioligand that is bound as $^*DR$ is trivial. In equation 3.16-3.19, this assumption permitted a simplification of the algebra by not substituting $[^*D]$ with $([^*D]_{\text{added}} - [^*DR])$. However, this restriction can be difficult to achieve experimentally; in these circumstances, the K_D value obtained is an *apparent* K_D value. Chang et al. (1975) demonstrated that when $[R]$ is not at least 10 fold lower than the K_D, the apparent K_D is a linear function of $[R]$. In circumstances where $[R]$ is not significantly less than K_D for $[D]$, the true K_D can be obtained by plotting $K_{D_{app}}$ versus reciprocal of the dilution factors of $[R]$, and extrapolating to the y intercept, i.e. when $[R] = 0$. Investigators can maintain sensitivity for detecting receptor biding in their assay while addressing the problem of too high a *concentration* of receptor by increasing the volume of the incubation (e.g. from 0.25 to 2.5 ml) such that the same number of ligand binding sites are present (i.e. sensitivity is not compromised) but $[R]$ has been reduced so that $K_{D_{app}}$ more closely approximates the true thermodynamic equilibrium dissociation constant, K_D.

Technical limitations that can confound data interpretation include, but are not limited to, the following examples:

1. If $^*D$ is not radiochemically pure, e.g. if the radioligand contains $^*D$ and D or multiply-labeled $^{**}D$, and if there is a difference in affinity constants for D, $^*D$, and/or $^{**}D$, then the data will not correspond to a single ligand interacting with a single receptor population via mass action. A specific example of radioligand heterogeneity is the use of racemic mixures as radioligands (Burgisser et al. [1981]).
2. If $^*D$ is degraded during the incubation so that equilibrium cannot be reacted, then the $K_{D_{app}}$ for $^*D$ will be a function of incubation duration. If suspected, the structural and biological integrity of $^*D$ at the end of the incubation should be documented.
3. If the $[R]$ changes during the incubation due to factors other than combinding with ligand, such as internalization of receptor in intact cell binding assays (cf. chapter 1), then K_D and B_{max} estimates will be in error.
4. If receptor inactivation occurs during an incubation and ligand binding stabilizes the receptor against degradation, then binding will not appear to

be described by a single, unchanging K_D for $^*D$ (in fact, apparent positive cooperativity of binding will result). The degree of stabilization will be lowest at low [$^*D$], and binding at these concentrations will more grossly underestimate binding which should be attained at equilibrium than with higher [$^*D$]. The presence of this technical limitation can be revealed by reincubating the receptor preparation with near-saturating [$^*D$] following the initial incubation at lower [$^*D$]. If no receptor degradation has occurred, then binding will equal that observed by incubating with the near-saturating [$^*D$] throughout.

Analyzing and Interpreting Saturation Binding Data

Figure 3-3 emphasizes that the relationship between receptor occupancy and radioligand concentration is a hyperbolic one when a saturable receptor population binding ligand in a freely reversible bimolecular reaction is being identified. Figure 3-4 is a schematic diagram of saturation binding data plotted two different ways. Figure 3-4A is a plot of [$^*DR$] versus [$^*D$], where [$^*D$] is plotted on a linear scale. Figure 3-4B is a plot of [$^*DR$] versus Log_{10}[$^*D$]. Figure 3-4A contrasts the two salient features of nonspecific versus specific binding.

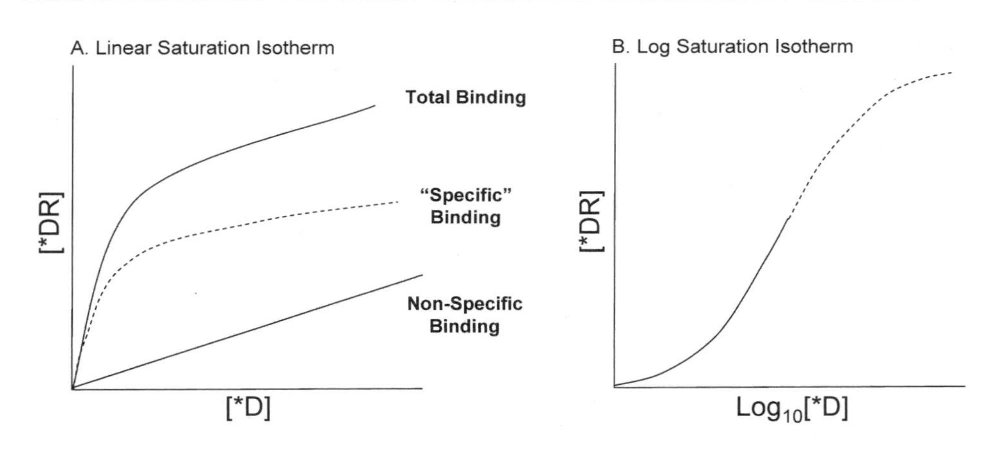

Figure 3-4. Differing representations of radioligand binding data. Experiments are performed where binding of $^*D$ to R is assessed as a function of increasing concentrations of radioligand $^*D$.

A. Linear saturation isotherm: the x axis is displayed as a *linear* function of increasing [$^*D$].

B. Log saturation isotherm: the x axis is displayed as a *logarithmic* function of increasing [$^*D$]. In this plot, a truncation of the isotherm due to the study of receptor binding over an insufficient concentration range of $^*D$ is readily apparent.

Nonspecific binding is a linear function of increasing radioligand concentrations, $[^*D]$, whereas specific binding (calculated as total binding minus nonspecific binding) appears to plateau. In reality, one can rarely saturate the receptor with the radioligand, despite the false impression given by the linear saturation isotherm that a plateau, or a rectangular hyperbola, has been obtained. Transforming the radioligand binding data to a plot of $[^*DR]$ versus $\text{Log}_{10}[^*D]$ (as in figure 3-4B) immediately discloses the truncation of the isotherm that occurs when binding studies are performed over too narrow a range of radioligand concentrations. Unfortunately, truncated isotherms are common because even *approaching* saturation of the receptor population often is experimentally impracticable. It is not just the cost of the radioligand that precludes using higher concentrations of $^*D$. Often a second, steeper slope of nonspecific binding is observed as the concentration of $^*D$ is increased. This second slope may represent the mixing of the radioligand into the membrane bilayer, since this phenomenon most commonly is observed for hydrophobic radioligands. Regardless of its origin, however, this second slope of nonspecific binding is worrisome, and investigators choose to study radioligand binding over a concentration range of $^*D$ where nonspecific binding demonstrates a single, shallow slope and represents a small fraction of total radioligand binding. As a consequence of the truncated binding isotherm obtainable with direct, experimental data, extrapolations of the data using non-linear regression analysis or of linear transformation is routine. What must be remembered, however, is that the validity of these extrapolations is directly related to the extent to which the ligand concentrations used reflect the entire isotherm (Klotz [1971] and [1982]).

Fitting Data to a Mathematical Model Using Non-Linear Regression

To determine the B_{max} and K_D for receptor binding to the radioligand, the data can be fit to equation 3.15 using non-linear regression. This analysis is based on several assumptions:

1. The binding follows the "law" (model) for mass action, described above.
2. There is only one population of receptors.
3. Only a small amount of the radioligand added binds to the receptor, so that the concentration of free radioligand is essentially identical to the concentration added.
4. There is no cooperativity, i.e. binding of a ligand to one binding site does not alter the affinity to another binding site. Thus, there is a single and unchanging K_D of the receptor for the radioligand.

Chapter 4 deals in considerable detail with models describing biological situations where multiple receptors or receptor affinity states exist, and how to analyze and interpret complex binding phenomena (see also Motulsky and Christopoulos [2003]). Despite the superiority of non-linear regression for data quantitation, there is considerable value in visual inspection of linear transformations of binding data. For example, if the data obtained do not correspond to a straight line when plotted by one of several linear transformations (assuming all technical limitations/artifacts have been eliminated), then the data do not correspond to a model of one ligand interacting with one receptor population with a single, unchanging K_D, and a more complex model (algebraic description) must be adopted for analysis of the data by non-linear regression.

Linear Transformations of Saturation Binding Data

Before the ready availability of computer-assisted analysis of binding profiles by non-linear regression, a *linear* transformation of binding data provided the advantage that three parameters could be easily determined: the slope of the line, its *y* intercept, and its *x* intercept. Three transformations (Scatchard, Rosenthal, and Hill) will be discussed below. The Lineweaver-Burk, or double reciprocal plot of $\frac{1}{[{}^*DR]}$ vs. $\frac{1}{[{}^*D]}$ will not be discussed further, since the data that most heavily influence the slope (and hence the intercept) of this plot are obtained at very low radioligand concentrations, where precision in estimating $[{}^*DR]$ is suspect.

Despite the limitations of linear transformations of binding data for rigorous determination of K_D and B_{max} values (Klotz [1971] and [1982]), they have significant utility for visualization and interpretation of data. Hence, their discussion here.

The Scatchard Plot

One of the most frequently used (and probably least understood) linear transformations of radioligand binding data is the so-called Scatchard plot. Its early popularity resulted from the potential ability to estimate receptor affinity from the slope of the plot and receptor density from the *x* intercept. Since receptor density ($[R]_{TOT}$) is the one parameter of ligand-receptor interactions that cannot be obtained from dose-response studies of receptor-mediated function (cf. chapter 2), a great deal of emphasis of radioligand binding studies has been on determining receptor density.

The form of the Scatchard plot that is applied most frequently to the analysis of radioligand binding data, and is shown in figure 3-5, is a plot of $[^*DR]/[^*D]$ on the ordinate versus $[^*DR]$ on the abscissa, or B/F versus B, where:

B	= Bound	= $[^*DR]$	=	concentration of ligand in the incubation present as ligand-receptor complex **at equilibrium.**
F	= Free	= $[^*D]$	=	concentration of free radioligand, $^*D$, present in the binding incubation **at equilibrium.**

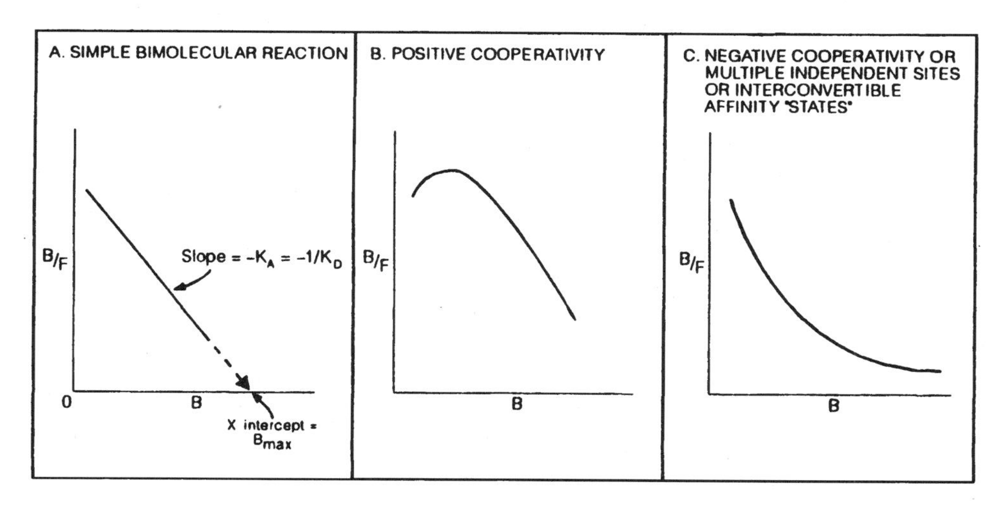

Figure 3-5. The commonly plotted Scatchard plot, B/F versus B, and its interpretation when nonlinear plots are observed. The Scatchard analysis, as originally described by Scatchard (1948), was a plot of $\overline{v}/c$ versus $\overline{v}$, where $\overline{v} = Y = B/B_{max}$ and $c = [^*D]$. In the original Scatchard analysis, the x intercept would = n, the number of ligand-binding sites per mole of receptor, R. In the Scatchard plot shown, the x intercept = $B_{max} = n[R]$.

The choice of B/F as the y axis and B as the x axis comes from algebraic rearrangement of equation 3.15 ($[DR] = [D][R]_{TOT}/(K_D + [D])$). This equation can be restated as:

$$B = \frac{(F)(B_{max})}{K_D + F} \qquad (3.16)$$

and rearranged to:

$$(B)(K_D) + (B)(F) = (F)(B_{max}) \quad (3.17)$$

dividing through by F:

$$(B/F)(K_D) + B = B_{max} \quad (3.18)$$

rearranging yields:

$$B/F = \frac{B_{max} - B}{K_D}, \text{ or } \frac{B_{max}}{K_D} - \frac{B}{K_D} \quad (3.19)$$

transformed to a linear expression, $y = mx + b$

$$B/F = -1/K_D \cdot B + B_{max}/K_D \quad (3.19)$$

Consequently, when a single ligand is interacting with a single population of receptors possessing a single affinity for the ligand, a plot of B/F versus B is a straight line and possesses a slope = $-1/K_D$ or $-K_A$. The x intercept ($y = 0$) is an estimate of B_{max} and the y intercept = B_{max}/K_D. A Scatchard plot for reversible binding to a single population of receptors possessing a single affinity for ligand is shown in figure 3-5A (the interpretation of figure 3-5B and 3-5C is discussed later).

Although a plot of B/F versus B typically is referred to as a Scatchard plot, it is *not* the plot that Scatchard derived (Scatchard [1949]; Munson and Rodbard [1983]). The genuine Scatchard plot derives from the relationship:

$$\bar{v}/c = k(n - \bar{v})$$

where

$$\bar{v} = \frac{\text{moles ligand bound}}{\text{moles of binding molecules}}$$

c = concentration of free ligand
k = equilibrium association constant
n = number of binding sites/mole of binding molecule

For comparison with earlier mathematical description, $\bar{v}$ is formally equivalent to Y in equation 3.8.

$c = [^*D]$ or $[D]$
$k = K_A$

and

$B_{\max} = n[R]_{TOT}$

A plot of $\bar{v}/c$ versus $\bar{v}$ gives a straight line if k is constant. Furthermore, for the binding determinations described by Scatchard, the concentration of the receptor species (moles/liter) would be *known*, and the x intercept of a plot of $\bar{v}/c$ versus $\bar{v}$ would give the value of n, the number of binding sites per mole of receptor, *not* the total density of binding sites, which is the value obtained by plotting B/F versus B.

The Rosenthal Plot

The Rosenthal plot was introduced to deal with situations where $[R]$ is an unknown, as typically is the case when binding to impure receptor preparations (Rosenthal [1967]). Consequently, the Rosenthal plot has been derived *assuming* that the concentration of receptors is an unknown, whereas the Scatchard analysis (above) assumes that the concentration of ligand-combining sites is known. The derivation of the Rosenthal plot is as follows:

$[b]$ = concentration of monovalent ligand bound to a macromolecule, M, with n equivalent and noninteracting sites
$[\mu]$ = concentration of free ligand
$[M]$ = concentration of the macromolecule, M, which behaves as the receptor
k = intrinsic association constant

$$k = \frac{[b]}{[\mu] \cdot [\text{free binding sites}]}$$

$$\frac{[b]}{[\mu]} = k \cdot [\text{free binding sites}]$$

If the concentration of free binding sites in the above expression is replaced by the difference of the concentration of available binding sites ($n \cdot M$) and the concentration of occupied binding sites ($[b]$), then

$$\frac{[b]}{[\mu]} = k\{(n \cdot [M]) - ([b])\} \tag{3.20}$$

which can be rearranged to

$$\frac{[b]}{[\mu]} = -k[b] + (n \cdot [M])k \tag{3.21}$$

If one plots $[b]/[\mu]$ versus $[b]$, then *both* coordinates are independent of the concentration of the macromolecule ($[M]$), which is rarely known. Equation 3.21 is a straight line with a slope $-k$, an intercept on the y axis of $k \cdot n \cdot [M]$ and an intercept on the x axis of $n \cdot [M]$. The product $n \cdot [M]$ is the number of available binding sites. If $[M]$ is known, n can be computed.

It can be seen that if one plots B/F versus B, as is usually done under the auspices of preparing a Scatchard plot, one is plotting the same raw data as $[b]/[\mu]$ versus $[b]$, since $[b] = B$ and $[\mu] = F$. What the Rosenthal analysis seeks to emphasize is that the interpretation of the x intercept is fundamentally different from that in a Scatchard plot. The x intercept is *not* a determination of $[R]_{TOT}$, but is a determination of the total number of binding sites, $n \cdot [R]_{TOT}$, and thus only estimates $[R]_{TOT}$ when $n = 1$. Since the value of n cannot be isolated out from the data, investigators typically determine B_{max} or $n \cdot [R]_{TOT}$ in units of fmol/mg membrane protein, thus focusing on the density of binding sites, and delay concerns about whether those binding sites represent $[R]$ or $n[R]$ until the receptor can be purified to homogeneity and the number of ligand-combining sites per receptor macromolecule can be determined directly.

Interpreting Scatchard (Rosenthal) Plots

When Scatchard (or Rosenthal) transformations are linear, the data are consistent with $^*D$ identifying a single population of receptors, R, with a single, unchanging affinity, K_D. The value of K_D can be obtained from the slope of the plot (as shown schematically in figure 3-5A for a Scatchard plot) and the concentration of binding sites (B_{max} or $n \cdot [R]$) can be determined from the intercept. Confidence in this interpretation is dependent on evaluating $^*DR$ over a sufficient range of $[^*D]$ to detect a second population of sites, if they exist. A plot of $^*DR$ versus $\log[^*D]$ (as in figure 3-4B) allows the investigator to determine if the data have been obtained over a sufficiently broad range of $^*D$ to reach an inflection point (Klotz [1982]).

Non-linear Scatchard plots can either be concave downward (figure 3-5B) or concave upward (figure 3-5C). The biological explanation for a concave downward (or upwardly convex) Scatchard plot, as in figure 3-5B, is that the receptor population demonstrates positive cooperativity, such that the affinity of the overall population increases with increasing receptor occupancy.

Positive cooperativity among receptors is rare, but is observed for some multisubunit receptors possessing >1 ligand binding site, such as the nicotinic acetylcholine receptor.

Most commonly, Scatchard plots that deviate from linearity are observed to be upward concave (downward convex) in shape, as shown in figure 3-5C. The biological interpretations of these findings include the possibilities that (1) there are multiple orders of noninteracting binding sites with unchanging and dissimilar affinities; (2) multiple "affinity states" of the receptor exist, for example, as a consequence of receptor-effector coupling, where a variable fraction of the total receptor population is coupled to the effector; or (3) negative cooperativity exists, such that the affinity of the overall receptor population decreases with increasing occupancy of the receptor.

Certain technical problems also can cause the artifactual appearance of upward concave Scatchard plots (Munson [1983]). Artifactual sources of negatively cooperative Scatchard plots include: (1) an inappropriate definition of non-specific binding; (2) the aggregation of the ligand at higher concentrations to a dimer or multimer that possesses a lower affinity for the receptor; and (3) a difference in the affinity or kinetic constants of $^{*}D$ and D, such that the receptor possesses a greater affinity for $^{*}D$ than for D, in a manner conversely analogous to the situation described above for artifactual positive cooperativity (see Taylor [1975]).

It is important to stress that heterogeneous steady state binding phenomena due to multiple, independent binding sites or to interconvertible affinity states always will resemble the situation of negative cooperativity, where the affinity of the receptor population is constantly decreasing as a consequence of increasing fractional occupancy of the receptor sites. The resemblance of steady state data manifested by binding to multiple receptor sites to data obtained from a system possessing negative cooperativity is a direct consequence of basic thermodynamics: when there are multiple receptor populations with differing affinities for ligand, the ligand will occupy the high-affinity population first, so that binding to the lower affinity population(s) will be detected only as ligand concentrations increase.

A non-linear Scatchard plot is just that: it is not a straight line. Often investigators resolve curvilinear Scatchard plots into two straight lines asymptotic to the steepest and most shallow portions of the curve. Apparent K_D values and receptor densities from these arbitrarily defined "two populations" of sites are then calculated. As discussed in detail in chapter 4, it is nonsense to take a curve and necessarily describe it by two straight lines. Furthermore, even when two independent receptor populations do underlie a curvilinear plot, it can be shown that the asymptotes of the steepest and most shallow portions do not accurately describe the affinity and density of these two receptor populations. In this case, independent biological data must be sought to reveal the molecular origin of the heterogeneity of ligand binding, and then computer-assisted analysis can be used to resolve that binding into its contributing components. Despite these limitations, the Scatchard or Rosenthal plot is still a good way to *display* the data because the plot reveals the non-homogeneity of $^*DR$ interactions, when they occur.

The Hill Plot

Another data transformation of saturation binding data is the so-called Hill plot. The equations introduced by Hill were an attempt to describe the observed, positively cooperative binding of O_2 to hemoglobin, a protein with four binding sites for O_2 located in four domains of the molecule. Hill (1910, 1913) based his equation on the model $R + nD \rightleftharpoons RD_n$, assuming that there is simultaneous binding to all of the sites, that no intermediate species exist, and that the forms of R in equilibrium are either the empty or the fully ligand-bound species (for hemoglobin, $n = 4$).

Hill equation:

$$Y = \frac{B}{B_{\max}} = \frac{[D]^n}{K_D + [D]^n} \text{ or } \frac{K_A[D]^n}{1 + K_A[D]^n} \tag{3.22}$$

When $n = 1$, the interaction of D with R obeys mass action for bimolecular interactions. When $n > 1$, the value of n obtained in this initial derivation was thought to represent the number of binding sites per molecule of R. However, this is now known not to be the case (see below).

Later models for the binding of O_2 to hemoglobin, such as those introduced by Adair (1925), assumed a *sequential* binding of O_2 to hemoglobin, such that intermediate species were considered to exist between the unliganded and fully liganded tetramer. If these derivations are generalized to ligand binding data, then the sequential binding of ligand D to receptor R can be expressed as:

$$[D] + [R] \overset{K_{D_1}}{\rightleftharpoons} [DR]$$

$$[D] + [DR] \overset{K_{D_2}}{\rightleftharpoons} [D_2R]$$

$$[D] + [D_2R] \overset{K_{D_3}}{\rightleftharpoons} [D_3R]$$

$$[D] + [D_{n-1}R] \overset{K_{D_n}}{\rightleftharpoons} [D_nR]$$

where

$$\text{Bound } (B) = \frac{B_{\max}[D]^n}{K_{D'} + [D]^n} \tag{3.23}$$

Equation 3.23 describes the net ligand binding isotherm, and $K_{D'}$ is a composite constant composed of the intrinsic dissociation constant K_D and interaction factors that determine the degree to which K_D is altered at each discrete binding step. As mentioned above, the value of n does *not* equal the number of binding sites for ligand, but is a more complex representation of the number of sites and the strength of the interaction among sites. The value of n is related to the number of sites only by the restriction that n cannot be greater than the number of sites (Monod et al. [1965]; Wyman and Gill [1990]; Weber [1992]). In fact, the observation that n does not equal the number of binding sites indicates an error in the Hill theory. Consequently, the Hill equation stands simply as an empirical description of complex binding phenomena.

The Hill equation can be transformed to a logarithmic form for convenience in plotting:

$$B \cdot K_{D'} + B \cdot [D]^n = B_{\max}[D]^n \tag{3.24}$$

$$B \cdot K_{D'} = (B_{\max} - B)[D]^n$$

$$\frac{B}{(B_{\max} - B)} \cdot K_{D'} = [D]^n$$

$$\frac{B}{(B_{\max} - B)} = \frac{[D]^n}{K_{D'}}$$

taking the log of both sides:

$$\log \frac{B}{(B_{\max} - B)} = n \log [D] - \log K_{D'} \tag{3.25}$$

Figure 3-6 is a schematic diagram of a Hill plot, where the y axis is $\log_{10} B/(B_{\max} - B)$ and the x axis is $\log [D]$, where $[D]$ is the concentration of *free* ligand in the incubation. The slope of the Hill plot, n, usually is referred to as the Hill coefficient, and denoted as n_H. The abscissa value where $\log B/(B_{\max} - B) = 0$ is the $K_{D'}$, since when $B = 1/2\ B_{\max}$, then $B/(B_{\max} - B) = 1$ and the log of

1 = 0. The K_D' value, as indicated above, is an *apparent* K_D that is a composite of the intrinsic dissociation constant, K_D, and the interaction factors that determine the degree to which the K_D is altered with the sequential binding of ligand.

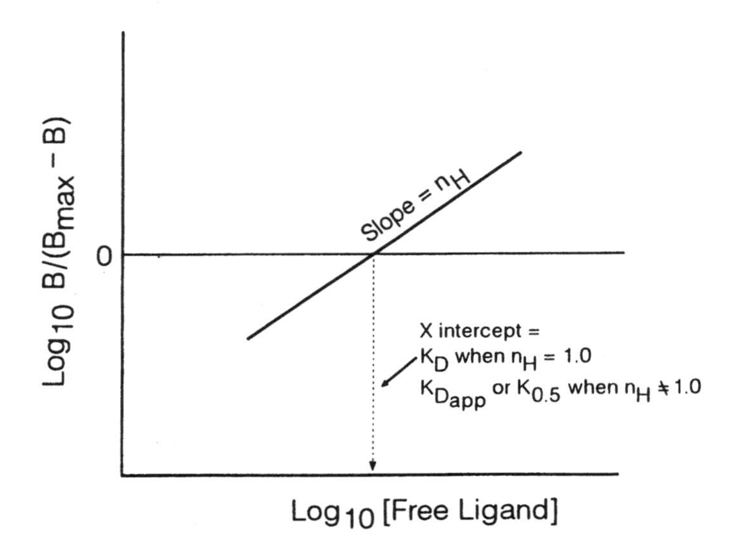

Figure 3-6. The Hill plot. The Hill plot also can be plotted as Log_{10} $Y/1 - Y$, where Y = fractional saturation = B/B_{max}.

This equation is formally the same as the four parameter logistic equation that will be discussed in more detail for analysis of binding data in chapter 4.

When $n_H = 1$, $K_{D'} = K_D$.

Interpretations of n_H:

$n_H = 1.0$: Ligand is binding to a single species of receptor via a simple, reversible bimolecular reaction driven by mass action. If multiple sites exist, they possess an identical affinity for D and are non-interacting.

$n_H > 1.0$: Positive cooperativity

$n_H < 1.0$: Interpretations include: negative cooperativity; multiple, non-interacting binding sites (R_1, R_2, R_3) for one ligand; multiple interconvertible affinity states.

One of the limitations of the Hill plot is that one should only consider data gathered over the range of approximately 30-70% occupancy in determining the slope, i.e., the n_H value. Data obtained at very low fractional occupancy and data obtained at near saturating occupancy interfere with a valid interpretation of n_H. The reason for not utilizing data points at low fractional occupancy (i.e., < 20-30%) is that cooperativity (either positive or negative) may not be "symmetrical" with respect to occupancy, i.e., a threshold occupancy might have to be attained before changes in affinity as a function of occupancy are perceived (Wyman [1948]; Cornish-Bowden and Koshland [1975]). The reason for not utilizing data points obtained at high fractional occupancy is that, in binding studies, one defines or calculates a B_{max} value, which is really only attained as $[D]$ approaches infinity. This setting of a B_{max} value means that as $B \rightarrow B_{max}$, $\frac{B}{B_{max} - B} \rightarrow \frac{B}{0}$ and, thus, $\frac{B}{0} \rightarrow \infty$, since division of a fraction by zero yields infinity. Consequently, at high receptor occupancy, the random error that occurs experimentally is considerably exaggerated, and the data transformation may become asymptotic with a vertical line. In fact, the transformation of B/B_{max} to $B/(B_{max} - B)$ results in considerable scale expansion at both ends of the Hill plot (Cornish-Bowden and Koshland [1975]).

Further Resolving Complex Binding Phenomena

Steady state binding data can reveal if binding represents a simple bimolecular reaction to a single population of receptors that possesses a single unchanging affinity for radioligand. However, steady state data *cannot* distinguish multiple, independent populations of receptors possessing unchanging affinities for ligands from negatively cooperative receptor populations whose affinity decreases as occupancy increases. Kinetic strategies, particularly those focusing on radioligand dissociation, are useful to distinguish between these molecular explanations for complex binding phenomena as in figure 3-5C. Quantitative analysis of complex binding data obtained from steady state experiments is the focus of chapter 4.

Determination of the Specificity of Radioligand Binding

The second criterion for binding expected for interaction of the radioligand with the physiologically relevant receptor is that the binding detected should demonstrate the appropriate specificity. The specificity with which the radioligand interacts with its binding sites is determined in competition binding studies, where the incubation of receptor-containing preparations with a constant concentration of the radioligand $^*D$ is carried out in the presence of increasing concentrations of unlabeled competitors *x, y* and *z*. As shown in figure 3-7A, data typically are plotted as a percentage of specific binding (total minus nonspecific binding) versus Log_{10}[competitor].

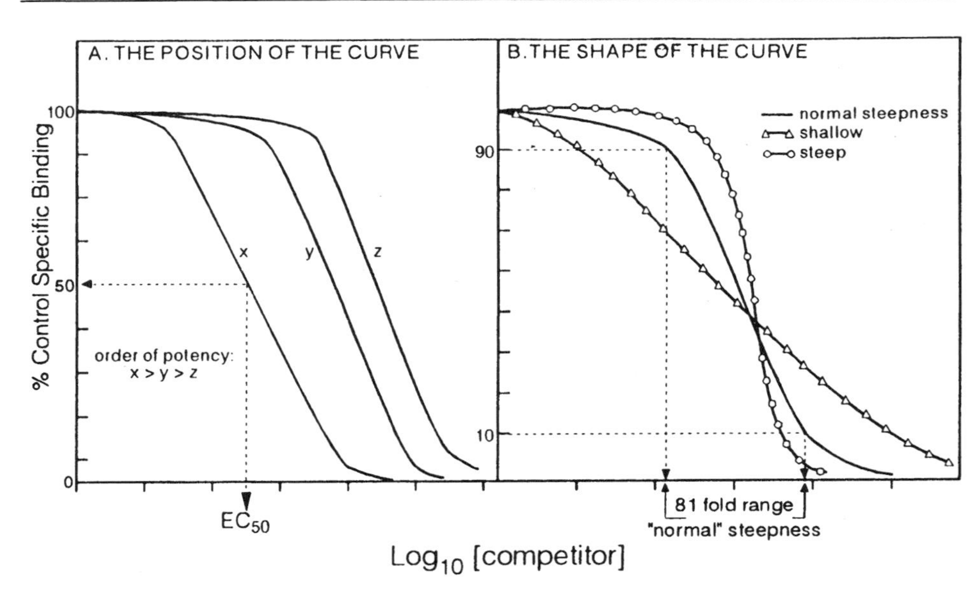

Figure 3-7. Competition binding profiles-a technique for assessing the specificity of the interaction between the radioligand and its binding site(s).
A. The relative potency of different unlabeled agents in competing for radioligand binding usually is expressed by calculating the EC_{50} value, i.e., the concentration of competitor that reduces specific radioligand binding detected in the absence of competitor (i.e., 100%), by half.
B. The shape of the competition profile. Deviation from "normal steepness," e.g., 10% to 90% competition over an 81-fold range of competitor, indicates a greater complexity of ligand-receptor interactions than accounted for by the reversible binding of a single ligand to a single population of receptors via mass action law.

When $^*D$ is interacting with the physiologically relevant receptor, the order of potency of unlabeled agents *x, y* and *z* in competing for binding to the receptor should parallel the order of potency of these agents in promoting (full agonists) or blocking (antagonists) the physiological effect(s) mediated via the putative receptor, *R*. The efficacy of an agonist (e.g. full vs. partial) or of an antagonist (null vs. inverse) may influence the comparison of these dose response curves quantitatively. Overall, however, the specificity reflected in regulating biological responses also should be reflected in the properties of these agents in competing for radioligand binding. Inherent in this discussion is the important principle that before a ligand binding site can be demonstrated to be a receptor of physiological interest, there must be a biological effect elicited by this ligand or its congeners to which the properties of radioligand binding can be compared.

Quantitation of the *Potency* of Competing Agents

To quantitate the potency of drugs in competing for the receptor, one determines from the competition binding curve the concentration of competitor that effectively competes for 50% of the specific radioligand binding (the EC_{50}). One can obtain the EC_{50} by logit-log analysis, so-called indirect Hill plots, computer-assisted nonlinear regression analysis, or (least accurately) by visual inspection. Indirect Hill plots plot the relationship:

$$\log \frac{[DR]_I}{[DR]-[DR]_I} = n \log[I] + n \log EC_{50} \qquad (3.26)$$

$[DR]$ = amount of binding in the absence of competitor, I
$[DR]_I$ = amount of binding in the presence of competitor, I
$[I]$ = concentration of competitor

The intercept on the abscissa is equal to the EC_{50}. When the slope of the line is -1, the data are consistent with both radioligand and competitor interacting with a single receptor population that possesses a discrete affinity for both ligands, i.e., neither cooperativity nor multiple receptor populations appear to exist. (Note that the negative sign of the slope is a consequence of plotting data as amount bound in the presence of competitor, and should be contrasted with the values plotted from equation 3.25, where the slope has a positive value. In addition, the n value in the $n \log EC_{50}$ expression for the x intercept arises because $K_{D'}$ or $K_{D_{app}} = EC_{50}{}^{n}$). A logit-log plot has certain similarities with the Hill plot (or indirect Hill plot), but in actuality is not mathematically equivalent or interconvertible with a Hill plot, as it is an empirical equation based on no biochemical models or thermodynamic parameters. The logit transform is defined as $\text{logit}(Y) = \log_e \left(\frac{Y}{1-Y}\right)$, where Y is a decimal fraction, i.e., $0 < Y < 1.0$.

The EC_{50} determined in competition binding studies is not equivalent to the K_D for the competitor, but depends on the concentration of the radioligand ($[^*D]$) present in the incubation. In certain situations, the K_D value for the competitor can be calculated from the EC_{50} value using the method introduced by Cheng and Prusoff (1973) and Chou (1974).

$$\mathrm{EC}_{50} = K_{D_I}\left(1+\frac{[{}^*D]}{K_{D{*}_D}}\right) \tag{3.27}$$

or

$$K_{D_I} = \frac{\mathrm{EC}_{50}}{\left(1+\frac{[{}^*D]}{K_{D{*}_D}}\right)} \tag{3.27A}$$

where K_{D_I} = equilibrium dissociation constant for competitor, I
$K_{D{*}_D}$ = equilibrium dissociation constant for radioligand, ${}^*D$

Thus, when $[{}^*D]$ is at its K_D', $\mathrm{EC}_{50} = 2 \times K_{D_I}$; when $[{}^*D]$ is present at trace concentrations ($*D \lll K_{D{*}_D}$), then $\mathrm{EC}_{50} = K_{D_I}$; when $[{}^*D]$ is present at $> K_D$, the deviation of EC_{50} from K_{D_I} is considerable. However, it must be emphasized that certain assumptions were made in the derivation of the Cheng and Prusoff/Chou equation that must be met by the experimental system in order for this calculation to be applied with validity. The criteria for valid application of this equation include:

1. ${}^*D$ must interact reversibly with a single population of R possessing a constant affinity for ${}^*D$ (i.e., the interaction obeys mass action law) *and* the competitor meets these same restrictions (i.e., the slope of an indirect Hill plot or a logit-log plot must equal - 1)
2. $[{}^*D]_{\text{added}} = [{}^*D]_{\text{free}}$
3. The concentration of the receptor is much less than the K_D for ${}^*D$ or for the competitor, such that the binding of ligand does not deplete the available radioligand or competitor present in the incubation (related to number 2, above)
4. The specific binding detected, after correction for nonspecific binding, is an accurate reflection of the amount of $[{}^*DR]$ formed
5. The incubation has proceeded long enough for steady state binding to be attained by the radioligand and *all* concentrations of the competitor.

As indicated in numbers 2 and 3, the use of the Cheng and Prusoff/Chou equation has assumed that the $[R] \ll K_D$ of the radioligand, such that $[{}^*D]_{\text{added}} = [{}^*D]_{\text{free}}$. However, this is often experimentally impractical if ${}^*DR$ complexes are to be quantifiable. Consequently, a method for calculating the K_D of the competitor from competition binding profiles when a significant fraction (>

5%) of the radioligand or competitor added is bound to the receptor has been proposed (Linden [1982]). In this calculation, the only additional piece of information that is needed beyond the values used for the Cheng and Prusoff/Chou calculation is the concentration of binding sites, $[R]_{TOT}$, which can be obtained by Scatchard or Rosenthal analysis or nonlinear curve-fitting algorithms (see chapter 4). The assumptions used to derive the relationships for the calculation below are violated, and the results are invalid, if the Scatchard or Rosenthal plots used to determine $[R]_{TOT}$ and K_D for the radioligand are not linear or if the value of K_I changes as a result of changing $[R]_{TOT}$ or $[^*D]$.

For the calculation below:

$[R]_{TOT}$ = total receptor, or binding site, concentration (= B_{max})
$[I]$ = concentration of competitor, I, free in the incubation at equilibrium in the absence of inhibitor, I
$[^*D]$ = concentration of radioligand, $^*D$, free in the incubation at equilibrium
K_D = K_D value for the radioligand, $^*D$
K_I = K_D value for the competitor, I
IC_{50} = IC_{50} of the inhibitor, I

To calculate the $[I]_{free}$ in the incubation:

$$[I] = IC_{50} - [R]_{TOT} + \frac{[R]_{TOT}}{2}\left[\left(\frac{[*D]}{K_D + [*D]}\right) + \left(\frac{K_D}{K_D + [*D] + [R]_{TOT}/2}\right)\right] \quad (3.28)$$

$[I]$ can then be substituted into the equation below to determine K_I:

$$K_I = \frac{[I]}{1 + \frac{[*D]}{K_D} + \frac{[R]_{TOT}}{K_D}\left[\frac{K_D + [*D]/2}{K_D + [*D]}\right]} \quad (3.29)$$

Equation 3.29 is formally equivalent to an earlier equation derived by Jacobs et al. (1975). However, it is worth emphasizing that the validity of the above calculation rests on the accuracy with which $[R]_{TOT}$ and $[^*D]_{free}$ can be determined.

Quantitation of the *Shape* of the Competition Binding Curve

Whether or not a competitor is interacting with the receptor via a simple bimolecular reaction or via greater complexity can be determined by scrutinizing the overall "shape" of the competition binding curve in a plot of Bound (or % Bound) versus Log_{10}[competitor]. When the radioligand $^*D$ and the competitor X interact reversibly with R via a simple bimolecular reaction, the competition curve for X will proceed from 10% to 90% competition over an 81-fold concentration range of competitor X (see figure 3-8B). This generalization was first introduced by Koshland (see 1970) for enzymes obeying Michaelis-Menten kinetics, but is described below in terms of $[D]$, K_D, and fractional saturation, Y.

$$Y = \frac{[D]}{K_D + [D]}$$

the values of Y at 90% and 10% saturation will be

$$0.9 = \frac{S_{0.9}}{K_D + S_{0.9}}$$

and

$$0.1 = \frac{S_{0.1}}{K_D + S_{0.1}}$$

Solving these simultaneously, one obtains:

$$\mathrm{R_s} = \text{cooperativity index} = \frac{S_{0.9}}{S_{0.1}} = \frac{\dfrac{0.9}{0.1}}{\dfrac{0.1K_D}{0.9K_D}} = 81$$

for positive cooperativity, $\mathrm{R_s} < 81$
for negative cooperativity or multiple orders of binding sites, $\mathrm{R_s} > 81$

Curves that proceed from 10% to 90% competition over an 81-fold concentration range of competitor are said to be of "normal steepness" and are characteristic of ligand-receptor interactions that describe a reversible bimolecular reaction that obeys mass action law. Curves proceeding from

10% to 90% competition over a greater than 81-fold range of competitor are frequently referred to as "shallow." Often, pseudo-Hill coefficients (pseudo-n_H) are calculated from competition binding curves using a plot as described in equation 3.26. A curve of normal steepness would possess a pseudo-n_H of 1.0; a shallow curve would possess a pseudo-$n_H < 1.0$. The prefix *pseudo-* for this calculated value of n_H emphasizes that a genuine interaction factor, n_H, for the competitor alone cannot be determined from competition binding studies, because multiple equilibria are occurring in the competition binding incubation:

$$*D + R \underset{k_2}{\overset{k_1}{\rightleftharpoons}} *DR$$

and

$$*I + R \underset{k_4}{\overset{k_3}{\rightleftharpoons}} IR$$

Consequently, some apparent deviation from simple mass action law is expected in applying the Hill equation to competition binding data.

Incubation duration is an important experimental parameter that must be considered in order for competition binding studies to yield an accurate EC_{50} value, and a non-misleading shape of the competition curve. Since the time required to achieve steady state binding for the radioligand is altered by the presence of the competitor (and vice versa for steady-state binding of the competitor) , the incubation duration must be sufficient for both competitor and radioligand to reach steady state occupancy of the receptor. The direction in which the EC_{50} shifts prior to equilibrium is primarily dependent on the rate constants for dissociation of the radioligand versus the competitor from the receptor (Aranyi [1980]; Ehlert et al. [1981]; Motulsky and Mahan [1984]). When the competitor and radioligand dissociate at the same rate ($k_2 = k_4$), as should occur when the competitor and radioligand have the same chemical structure, the EC_{50} for the competitor decreases over the course of the incubation, i.e., the competition curve continually shifts to the left until it reaches its equilibrium position. The same situation is true when the competitor dissociates more slowly than the radioligand ($k_4 < k_2$), that is, the competition curve shifts to the left prior to reaching its equilibrium position. In contrast, when the competitor dissociates from the receptor more quickly than the radioligand, the EC_{50} first decreases and then increases until it reaches its equilibrium position. In this situation, it can be shown that the minimum (leftmost) value of the EC_{50} will be the K_I, whereas in all other cases the EC_{50} will be greater than the K_I (Chou [1974]; Motulsky and Mahan [1984]). It should be emphasized that the early decrease in EC_{50} may occur quickly and thus go unnoticed by the investigator in empirical studies of time-dependent changes in EC_{50} with incubation duration (Ehlert et al. [1981]).

The ability to predict the time to attain equilibrium for a competition binding study using computer modeling techniques has provided certain useful "rules of thumb" for setting the duration of a binding incubation:

If $k_2 << k_4$, meaning that the competitor dissociates from the receptor more rapidly than the radioligand, then (1) at very low $[^*D]$, i.e., $[^*D] << K_D$, equilibrium is achieved at $3.5/k_2$. Since, for dissociation of a bimolecular complex, $k_2 = 0.693/t_{1/2}$, then equilibrium is achieved at approximately five times the $t_{1/2}$ for dissociation of the radioligand; (2) at very high $[^*D]$, i.e., $[^*D] >> K_D$ for $^*D$, equilibrium is achieved at $1.75/k_2$, i.e., only twice as fast (see Motulsky and Mahan [1984).

Alternatively, if $k_2 >> k_4$, meaning that the radioligand dissociates significantly more rapidly than the competitor, then the concentration of the radioligand added is irrelevant in terms of the duration of incubation needed to reach equilibrium, and equilibrium can be shown to be reached in $1.75/k_4$.

Assessment of the effect of incubation duration on the characteristics of competition binding profiles using computer modeling techniques also has demonstrated that the *shape* of the competition profile changes with time (Motulsky and Mahan [1984]). With simulations using a variety of kinetic constants, it has been shown that, prior to equilibrium, the slopes (calculated around the EC_{50} of each curve) are always between 1.0 and 1.3 for curves that at equilibrium have a slope of 1.0. This is a useful piece of information that is not necessarily intuitively obvious. For example, a number of investigators have observed a time-dependent increase in the EC_{50} for agonists, but not antagonists, in competing for beta-adrenergic receptors on intact cells. (For examples, see Pittman and Molinoff [1980] and Insel et al. [1983].) These data have been interpreted as a manifestation of agonist-induced desensitization. However, an increase in the EC_{50} as a function of incubation duration would be completely consistent with the kinetics of competitive inhibition if the dissociation rate constant of the competing agonist (k_4) were greater than that of the radiolabeled antagonist (k_2) in these studies. However, the reported changes in agonist binding properties include a time-dependent change in the shape of the competition binding curve, such that the slope factor for agonist competition profiles is <1.0 at early time points and increases to 1.0 at equilibrium. This latter observation cannot be accounted for by simple competitive binding theory and suggests that the anomalous behavior of agonist binding observed in these kinetic experiments may be a reflection of an agonist-induced molecular event of mechanistic interest.

Assessment of B_{max} (as well as K_D) Values from Homologous Competition Binding Curves

Where a competition binding experiment is performed using the same compound as both radioligand and competitor, the curve can be described as a "homologous" competition binding curve. These curves are a special case of competition binding studies where the data can be used to define not only the affinity of the ligand, but also receptor density. To analyze a curve with this in mind, certain assumptions (reminiscent of those in analysis of saturation binding data using increasing concentrations of radioligand, as described earlier) must be met:

1. The receptor has identical affinity for the labeled radioligand and unlabeled radioligand and unlabeled competitor. Since iodination can change the binding properties of ligands, it may be wise to use an iodinated, but non-radioactive, ligand as the competitor (e.g. ^{127}I-ligand).
2. There is no evidence of complexity in binding, i.e. curves are of normal steepness.
3. No ligand depletion occurs during the incubation, such that [ligand] added = [ligand] free.

A homologous competition curve is analyzed using the same equation used for a one-site plot of "heterologous" competition binding curves, i.e. when the competitor is structurally distinct from the radioligand, as described in equation 3.27. The K_I can be determined in a homologous competition binding experiment by assuming that the radioligand and competing ligand have the same affinities, so that $K_D = K_I$. This allows a simplification of the Cheng and Prusoff/Chou equation (3.27) to:

$$K_I = \mathrm{EC}_{50} - [{*D}] \tag{3.30}$$

To determine the B_{max}, the specific binding, B (expressed in units of pmol/mg) is divided by the fractional occupancy, $B/B_{\max}$, calculated from the $K_{D_{*D}}$ and $[{*D}]$:

$$\frac{B}{B/B_{\max}} = B_{\max} = \frac{\text{specific binding}}{\text{fractional occupancy}} = \frac{\text{specific binding}}{\dfrac{[{*D}]}{K_D + [{*D}]}} \tag{3.31}$$

Properties of Allosteric Modifiers as Manifest in "Competition" Binding Studies

Occassionally, agents evaluated in competition binding assays do not interact at the orthosteric binding site and, hence, are not strictly "competitive." When an allosteric modifier evokes positive cooperativity, an increase in radioligand binding occurs as increasing concentrations of the agent are added to the incubation, although a "ceiling" ultimately is reached. In contrast, when allosteric modulators have significant negatively cooperative effects, they decrease radioligand binding, and this decrease in binding of trace radioligand concentrations may be mistaken for competition. Whereas negative allosteric

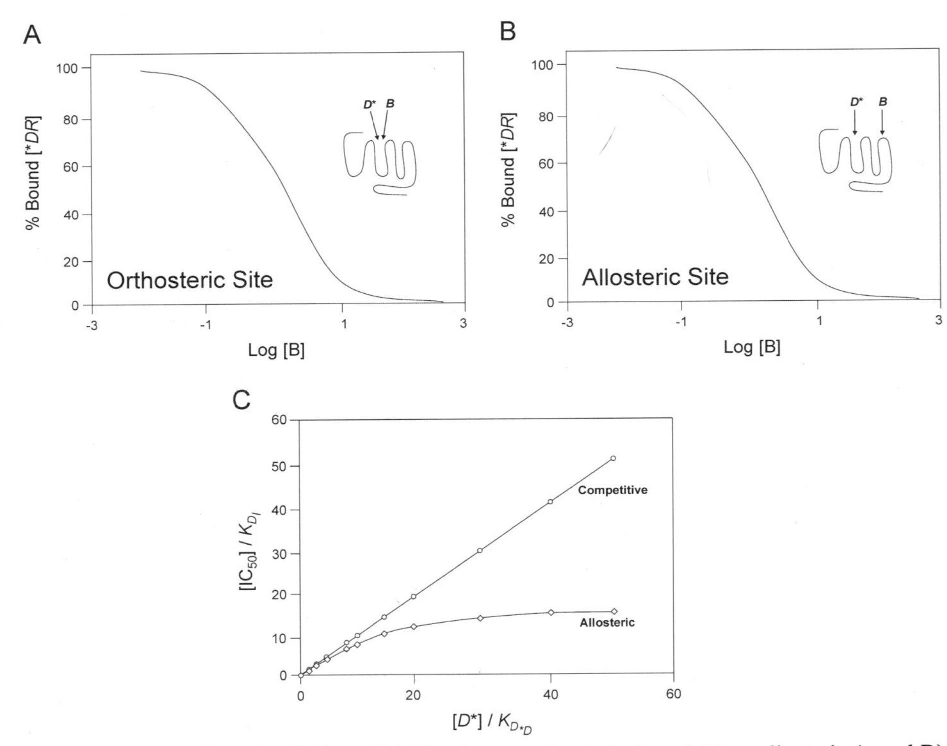

Figure 3-8. Inhibition of radioligand binding by an orthosteric (panel A) or allosteric (panel B) inhibitor, denoted as *B*. Panel C provides a diagnostic assessment to differentiate between competitive versus allosteric antagonism by agent *B* of binding of the radioligand, **D*, i.e. to compare the EC_{50} value for competition for the radioligand as a function of radioligand concentration. In the example shown, the value of α for the allosteric modulator as modeled is α=0.1. See text for further discussion. Figures modified from Kenakin [2001].

modifiers may deceptively look like competitors of radioligand binding, the allosteric nature of an agent's effect on receptor affinity can be revealed in independent kinetic experiments (see later text related to figure 3-8)

The scheme below can be used to describe the effects of allosteric agents on monomeric receptors Christopoulos and Kenakin [2002]):

$$\begin{array}{ccc} \overset{I}{\underset{+}{}} & & \overset{I}{\underset{+}{}} \\ {}^{*}D + R & \overset{K_{D*_D}}{\rightleftarrows} & {}^{*}DR \\ K_{D_I} \Updownarrow & & \Updownarrow \alpha K_{D_I} \\ {}^{*}D + RI & \overset{\alpha K_{D*_D}}{\rightleftarrows} & {}^{*}DRI \end{array}$$

Here, $^{*}D$ is defined as the radioligand interacting at the orthosteric site, and I as the ligand at the allosteric site on the same monomeric receptor, R. (This model is formally indistinguishable from the ternary complex model for G protein-coupled receptors, where I = the heterotrimeric G protein. The ternary complex model and its application to G protein-coupled receptor systems is discussed extensively in chapter 4.)

When α = 0, then there is no allosteric activity, and I, like $^{*}D$, is a competitor at the orthosteric site. When the value of α is less than 1, apparent negative cooperativity exists, when α is greater than 1, positive cooperativity of radioligand binding is occurring as a consequence of binding the allosteric modulator. A value of $\alpha = 1$ would characterize an allosteric interaction that does not alter orthosteric ligand affinity at equilibrium.

From previous discussions, the properties of competitive agents in radioligand binding studies are straightforward. A competitive ligand interacting at the orthosteric site to which radioligand binds ultimately decreases radioligand binding to levels defined as non-specific binding. The relationship between the concentration of competitor required to reduce a defined level of specific radioligand binding to 50%, i.e. the EC_{50}, is given by the Cheng and Prusoff/Chou equation (3.27), i.e.

$$\mathrm{EC}_{50} = K_{D_I}\left(1 + \frac{[{}^{*}D]}{K_{D*_D}}\right)$$

Stated another way, the [I] required to reduce a defined level of specific radioligand binding to 50% can be calculated as:

$$\frac{[I]}{K_{D_I}} = \frac{[^*D]}{K_{D^*_D}} + 1 \tag{3.32}$$

According to this relationship, the concentration of inhibitor/competitor in a radioligand binding assay, expressed as a multiple of K_{D_I}, is linearly related to the concentration of radioligand in the assay ($\frac{[I]}{K_{D_I}} = \frac{1}{K_{D^*_D}} \cdot [^*D] + 1$, or $y = mx + b$).

Christopoulos and Kenakin (2002) have derived a corresponding relationship for allosteric ligands:

$$\mathrm{EC}_{50} = K_{D_I} \left[\frac{[^*D] + K_{D^*_D}}{\alpha [^*D] + K_{D^*_D}} \right] \tag{3.33}$$

where EC_{50} represents the concentration of radioligand binding inhibitor, I, that decreases specific radioligand binding by 50%. When the concentration of radioligand is low, relative to its K_D value ($^*D <<< K_{D^*_D}$), then the EC_{50} will $\cong$ the K_{D_I}. However, the most important consequence of this algebraic description is that the relationship between $\frac{[I]}{K_{D_I}}$ and $[^*D]$ is not a linear but rather a hyperbolic one. Thus, as shown in figure 3-8C, a useful way to differentiate competitive from allosteric antagonism of radioligand binding is to compare the EC_{50} value of a presumed competing agent as a function of radioligand concentration. A plot of $\frac{\mathrm{EC}_{50}}{K_{D_I}}$ versus $\frac{[^*D]}{K_{D^*_D}}$ will be linear when the inhibitor, I, is competitive, and will be curvilinear when I is allosteric in nature. Another prediction is that the ability of an allosteric inhibitor to decrease radioligand binding to non-specific binding levels will depend on the magnitude of α, the cooperativity factor, reflecting the inability of an allosteric modifier to produce a significant enough decrease in radioligand affinity to bring the signal to non-specific radioligand binding levels. Hence, an inhibition curve where radioligand binding is not inhibited to non-specific binding may suggest that the inhibitor is allosteric in nature, warranting more rigorous analysis, using kinetic strategies discussed below.

DETERMINATION OF RATE CONSTANTS FOR RADIOLIGAND ASSOCIATION AND DISSOCIATION

In addition to determining the saturability and selectivity of ligand binding, a third criterion expected for the binding of a radiolabeled hormone or drug ($^*D$) to the physiologically relevant receptor is that the time course of binding should correspond to, or precede, the time course characteristic of the physiological effect elicited by $^*D$. To quantitate the rate of binding of $^*D$ to its binding site, the rate of radioligand association to and dissociation from the putative receptor is determined.

To reiterate definitions, the binding reaction of interest is:

$$*D + R \underset{k_2}{\overset{k_1}{\rightleftharpoons}} *DR$$

where k_1 = association rate constant
k_2 = dissociation rate constant

Determination of the Association Rate Constant

The rate of formation of $^*DR$ over time can be expressed as:

$$\frac{d[*DR]}{dt} = k_1[*D][R] - k_2[*DR] \tag{3.34}$$

At equilibrium, the rate of formation of $^*DR$ equals the rate of dissociation of the $^*DR$ complex, i.e., $d[^*DR]/dt = 0$.

Under circumstances where $[^*D]$ is added at concentrations in considerable excess of $[R]$, the $[^*D]$ can be assumed, as a first simplification, not to change throughout the incubation, whereas $[R]$ decreases considerably as $[^*DR]$ increases. Thus, the rate of association is described as "pseudo first-order" where k_1' = pseudo first-order rate constant = $k_1[R]$. Association data are obtained and plotted as the quantity of bound ligand, $[^*DR]$, obtained versus time, t (figure 3-9A). The reaction continues until equilibrium is reached, i.e., $d[^*DR]/dt = 0$ and an equilibrium value of binding $[^*DR]_{eq}$ is attained.

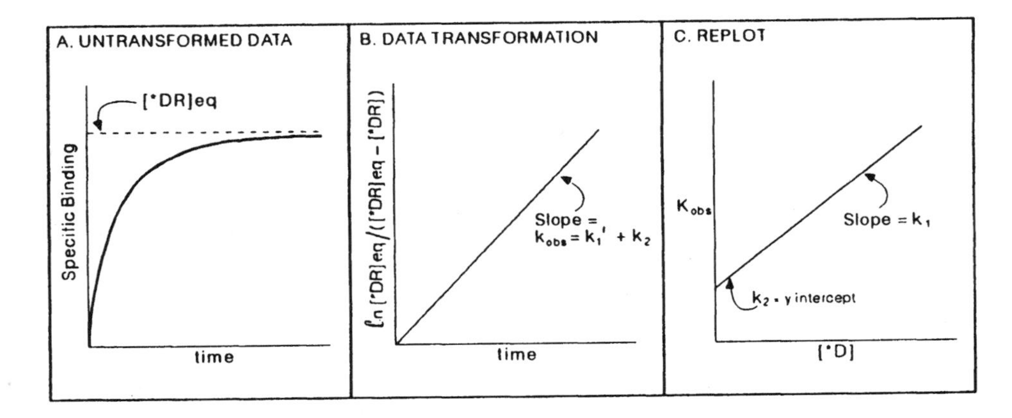

Figure 3-9. Determination of the association rate constant for the formation of a $^*DR$ complex.
A. Untransformed data. The amount of $^*DR$ formed is plotted versus time. At equilibrium, no increment in binding is detected, and a value for $[DR]_{eq}$ can be estimated.
B. Calculation of the observed rate constant, $k_{obs} = (k_1[D] + k_2) = (k_1' + k_2)$.
C. Replot to determine the association rate constant, k_1, from the k_{obs} values obtained at varying concentrations of radioligand $^*D$.

At any time, t: $[R] = [R]_{TOT} - [^*DR]_t$

At equilibrium, the net formation of $^*DR = 0$, such that $\dfrac{d[*DR]}{dt} = 0$

Substituting the above relationship for R, and solving for equation 3.34 at equilibrium yields:

$$k_1'([R]_{TOT} - [*DR]_{eq}) = k_2[*DR]_{eq} \tag{3.35}$$

$$k_2 = \frac{k_1'\left(\left[R\right]_{TOT} - \left[*DR\right]\right)}{\left[*DR\right]_{eq}}$$

Substituting the above relationship for k_2 into equation 3.34 at any time, t, yields:

$$\frac{d\left[*DR\right]}{dt} = k_1'\left(\left[R\right]_{TOT} - \left[*DR\right]\right) - \left[k_1' \frac{\left[R\right]_{TOT}}{\left[*DR\right]_{eq}} - k_1'\right]\left[*DR\right]$$

which reduces to:

$$\frac{d[*DR]}{dt} = k_1'[R]_{TOT}\frac{\left([*DR]_{eq} - [*DR]\right)}{[*DR]_{eq}}$$

which rearranges to:

$$\frac{d[*DR]}{\left([*DR]_{eq} - [*DR]\right)} = k_1'\frac{[R]_{TOT}}{[*DR]_{eq}} \cdot dt$$

Integrating this equation from t_1 to t_2 yields:

$$ln\frac{\left([*DR]_{eq} - [*DR]_{t_1}\right)}{\left([*DR]_{eq} - [*DR]_{t_2}\right)} = \frac{k_1'[R]_{TOT}}{[*DR]_{eq}}(t_2 - t_1)$$

This equation is general, and can be used for any two time points on an association curve. This equation can be further reduced by setting $t_1 = 0$. Therefore, $[^*DR]t_1 = 0$.

$$ln\frac{[*DR]_{eq}}{\left([*DR]_{eq} - [*DR]_t\right)} = \frac{k_1'[R]_{TOT}}{[*DR]_{eq}} \cdot t$$

shown above:

$$k_2 + k_1' = \frac{k_1'[R]_{TOT}}{[*DR]_{eq}}$$

and

$$k_1' = k_1[*D]$$

$$ln\frac{[*DR]_{eq}}{\left([*DR]_{eq} - [*DR]\right)} = \left(k_2 + k_1[*D]\right)t \tag{3.36}$$

The observed rate constant, k_{obs}, is defined as $(k_1[D] + k_2) = (k_1' + k_2)$.

When $ln[^*DR]_{eq}/([^*DR]_{eq} - [^*DR])$ is plotted versus time, the plot yields a straight line that passes through the *0,0* intercept and possesses a slope of $k_{obs} = (k_2 + k_1[^*D]) = (k_2 + k_1')$ (see figure 3-9B). The units of k_{obs} are min^{-1}. There are two alternative ways to determine the bimolecular association rate constant, k_1 (min^{-1} M^{-1}) from the k_{obs} value. First, since $k_{obs} = k_1[^*D] + k_2$, and the dissociation rate constant, k_2, can be determined independently (see below), then k_1 can be obtained by calculation: $k_1 = (k_{obs} - k_2)/[^*D]$. Alternatively, the k_{obs} can be determined at several $[^*D]$, and the k_{obs} plotted versus $[^*D]$ to yield a straight line whose y intercept equals k_2 and whose slope equals k_1 (cf. figure 3-9C). This latter method for estimating value for k_1 is more rigorous, and generally offers the investigator more opportunities to detect internal inconsistencies within the data.

Interpretation of association rate data requires a recollection of the assumptions inherent in the derivation:

1. The association of D with R is a reversible bimolecular reaction driven by the law of mass action.
2. The $[^*D] >> [R]$, and thus it can be assumed that $[^*D]_{added} = [^*D]_{free}$ and that a pseudo first-order reaction can be assumed to approximate the initial rate of association.
3. The measured $[^*DR]$ accurately reflects the concentration of this species, i.e., bound is effectively separated from free radioligand without dissociation of the $^*DR$ complex. Furthermore, non-specific binding of $^*D$ is defined with validity.

When these assumptions are met, deviation from linearity of a plot of the association rate data as in figure 3-9B most often suggests a greater complexity than the interaction of $^*D$ with a single population of R possessing a fixed affinity for $^*D$.

Determination of the Dissociation Rate Constant

To determine the rate of radioligand dissociation from the receptor, the experimental conditions are "fixed" so that the association of $^*D$ and R are negligible, and only dissociation can be measured:

$$\frac{d[*DR]}{dt} = -k_2[*DR]$$

Experimentally, association is made negligible by either (1) "infinitely" diluting an equilibrated solution of $^*D$ and R so that further binding of radioligands or rebinding of radioligand, once dissociated, cannot be detected, or (2) by adding a large excess of unlabeled competing ligand (D) so that subsequent to dissociation of $^*D$ from the receptor, rebinding of the radioligand cannot occur, due to competition of the excess of unlabeled D for the rebinding of dissociated $^*D$. Often it is useful to compare the observations obtained using both of these approaches (see figure 3-6 and discussion below).

Rearrangement and integration of $\frac{d[*DR]}{dt} = -k_2[*DR]$ gives

$$ln\frac{[*DR]}{[*DR]_0} = -k_2t \qquad (3.37)$$

$[^*DR]_0$ = $[^*DR]$ bound at time = 0, i.e., time the dissociation phase was initiated by infinite dilution or addition of excess unlabeled radioligand.

$[^*DR]$ = $[^*DR]$ bound at any time, t

For first-order reactions, such as occur when a single bimolecular species, $^*DR$, dissociates, the rate constant k_2 can be shown to be related to the $t_{1/2}$ in a constant fashion: $k_2 = 0.693/t_{1/2}$, where $t_{1/2}$ equals the time it takes for half of the $^*DR$ complexes to dissociate.

A plot of $ln[^*DR]/[^*DR]_0$ versus time yields a linear transformation with a slope of $-k_2$. When the data are plotted as the $\log_{10}$ of $[^*DR]/[^*DR]_0$ versus time (figure 3-10A), rather than the natural log (ln) of the ratio of $[^*DR]/[^*DR]$, then the slope of the line obtained is $-2.303\ k_2$. The units of k_2 are min^{-1}. Again, the assumptions inherent in the mathematical derivation of the plot dictate its interpretation. The observation of a linear $\log_{10}$ $[^*DR]/[^*DR]_0$ versus time plot is consistent with first-order dissociation, i.e., the kinetic profile expected for the dissociation of a single bimolecular complex, $^*DR$, which possesses a single and unchanging affinity for $^*D$ (cf. figure 3-10A).

When binding of $^*D$ to R represents binding to a single class of receptors binding $^*D$ in a reversible fashion with a fixed affinity, then it is expected that the K_D determined for the receptor from steady state (saturation) binding data

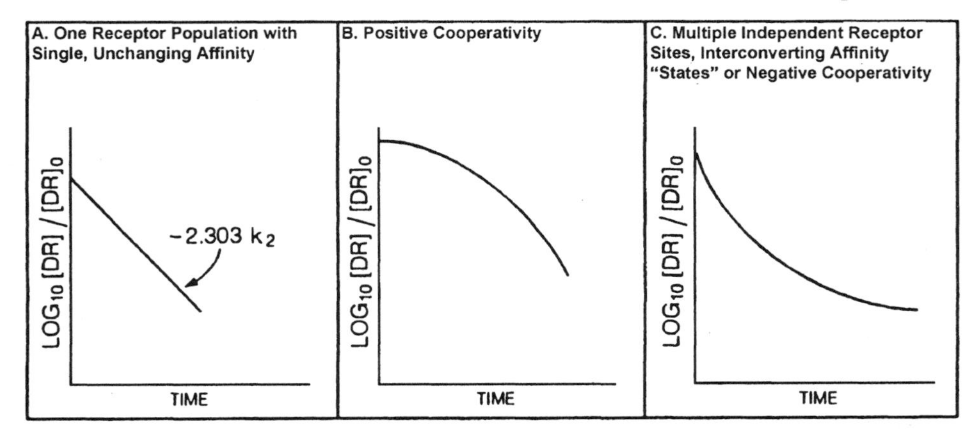

Figure 3-10. Dissociation of $^*DR$ complexes as a function of time.
A. A straight line on this plot is what is expected for a first-order reaction, and is consistent with what is expected for dissociation of radioligand from $^*DR$ when $^*DR$ is formed via a bimolecular reaction: $*D + R \rightleftharpoons *DR$. In this situation, one expects the dissociation of a single species, $^*DR$, with a single rate constant k_2 (min^{-1}).
B. Dissociation profile characteristic of a positively cooperative system when the dissociation phase is initiated by "infinite" dilution of the incubation so that rebinding of radioligand cannot occur.
C. When the dissociation phase is initiated by infinite dilution of the incubation, this dissociation profile would be characteristic of (1) a negatively cooperative system, (2) a system possessing multiple independent populations of sites, or (3) a system possessing multiple receptor affinity states, as examples. Contrast this with what is expected for a negatively cooperative binding system when dissociation is measured in the presence of excess unlabeled ligand to prevent rebinding of radioligand (figure 3-6).

---

should be equivalent to the K_D calculated from kinetic data:

$$K_D = \frac{k_2 \ \text{min}^{-1}}{k_1, \ \text{min}^{-1}\,\text{M}^{-1}} = \text{Molar}$$

A ratio of k_2/k_1 that does *not* equal (within reasonable experimental error) the value for K_D obtained using equilibrium binding studies may indicate an interesting biological phenomenon is occurring, such as a ligand-induced, time-dependent changes in receptor affinity paralleled by time-dependent changes in values for K_D (or $K_{D_{apparent}}$). Ligand- and time-dependent changes in receptor affinity might be due to changes in receptor conformation, ligand-fostered receptor association with effector proteins, or to ligand-induced covalent modification of the receptor. Each of these possibilities can be tested using independent, and complementary, biochemical strategies. Thus, kinetic analyses can be viewed as a means either to re-confirm estimates of binding parameters obtained from equilibrium binding data or to reveal interesting, biologically important changes in receptor properties that are ligand- and time-dependent.

Deviation from linearity suggests the existence of multiple binding sites or cooperativity. Figure 3-10B is a schematic diagram of what is expected for dissociation from a positively cooperative system. Since the affinity of the receptor population decreases as occupancy decreases, the rate of dissociation continually accelerates to a limit rate, which is characteristic of the intrinsic rate of dissociation from binding sites unperturbed by occupancy of other receptor sites. Figure 3-10C is a schematic diagram of what is expected for multiple populations or affinity states of binding sites or, alternatively, negative cooperativity among the binding sites. At early times, the rate of dissociation is more rapid, since the rapidly dissociating lower affinity complexes dissociate first and, in the case of negative cooperativity, the overall affinity of the population increases as occupancy decreases. It must be emphasized that deviations from linearity due to the existence of cooperative receptor systems, i.e., systems where $K_{D_{\mathrm{app}}}$ is changing continually as a function of the fraction of receptors occupied, are detectable only when (1) the radioligand occupies a fraction of receptors sufficient to induce the cooperative effect in the association phase of the experiment and (2) dissociation is monitored by infinitely diluting the incubation volume, not by adding excess unlabeled ligand.

Dissociation Strategies to Distinquish Negative Cooperativity from Multiple, Independent Receptor Populations

When excess unlabeled ligand is added in a cooperative system as a means to eliminate further association and thus focus solely on dissociation, a plot of $\log_{10}[^{*}DR]/[DR]_0$ versus time should be a straight line. This is because a receptor population that exhibits cooperative behavior nonetheless is characterized by a single $K_{D_{app}}$ value at saturating receptor occupancy. The understanding that the rate of dissociation of a cooperative system depends on the extent of receptor occupancy of that system is the basis for the kinetic approach developed by DeMeyts (1976) to assess the possibility of negative cooperativity among radioligand binding sites. The experimental design is shown schematically in figure 3-11.

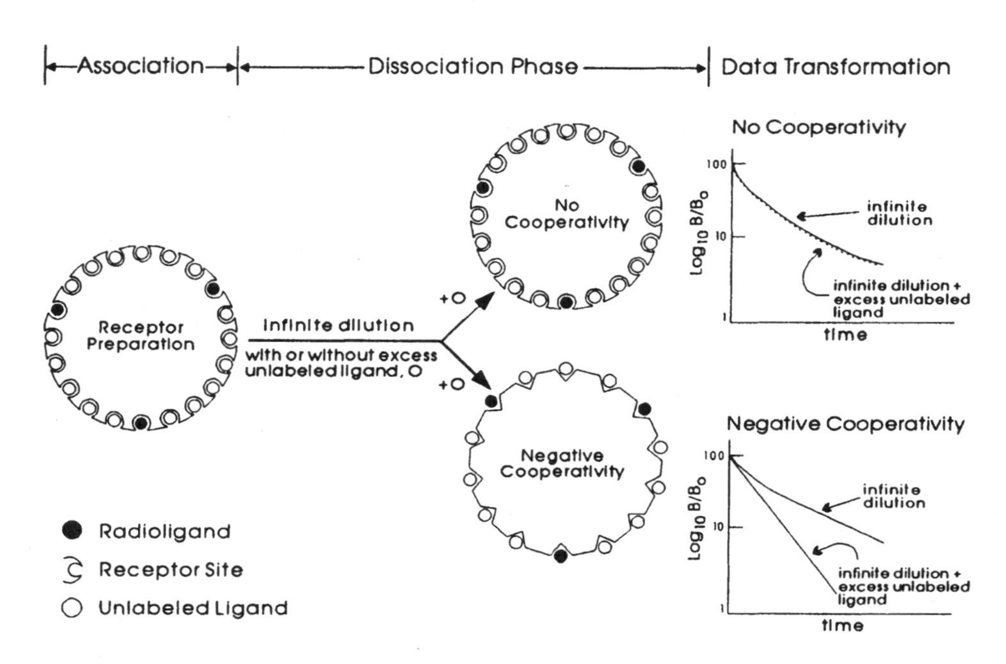

Figure 3-11. A kinetic protocol for evaluating the possible existence of negative cooperativity among receptor binding sites.

Radioligand at a concentration occupying only a small fraction of the overall receptor population is incubated with the biological preparation of interest until binding reaches steady state. This association phase is terminated by "infinite dilution" of the incubation. Infinite dilution prevents further *association* of the radioligand with the receptor because the decreased concentration of the reactants, $^*D$ and R, decreases the probability of $^*DR$ interaction to virtually zero. Infinite dilution is determined empirically (usually 100-fold is sufficient), and is operationally defined as a dilution sufficient to prevent rebinding of the ligand, once dissociated.

To assess possible existence of negative cooperativity among the radioligand binding sites, the dissociation phase of the experiment is monitored under two conditions: "infinte dilution" and "infinite dilution" plus the addition of excess unlabeled ligand at a concentration sufficient to occupy all of the binding sites. In this arm of the experiment, the binding pocket of the receptor population is fully occupied, either by radioligand or by the unlabeled ligand that interacts with the same population of recognition sites as the radioligand added during the dissociation phase. If the receptor sites bind ligand independently of one another, then filling the unoccupied receptors with excess unlabeled ligand during the dissociation phase should not alter the rate of dissociation of the radioligand. In contrast, if negative cooperativity exists, filling receptors with unlabeled agent will decrease the overall affinity of the receptor population and accelerate the rate of radioligand dissociation. This is because $K_D = k_{off}/k_{on}$, and a decrease in affinity is paralleled by an increase in K_D. Since the k_{on} is often invariant, limited only by the rate of ligand diffusion to the receptor binding site, then an increase in K_D will almost always be associated with an increased rate of dissociation, or k_{off} (DeLean and Rodbard [1979]; DeMeyts [1976]). Consequently, a comparison of radioligand dissociation under conditions of infinite dilution versus infinite dilution plus excess unlabeled ligand provides a diagnostic strategy for determining whether steady state binding data that do not conform to a simple biomolecular reaction are due to multiple, independent receptor sites or negative cooperativity among the receptor binding sites. The $K_{D_{app}}$ does not change as a function of receptor occupancy when the ligand binding sites of multiple receptor populations bind independently of one another, whereas dissociation rates accelerate with increasing occupancy of negatively cooperative receptor systems.

Revealing Allosteric Modulation of Receptor Binding Properties Using Kinetic Strategies

As for evaluation of negatively cooperative systems, studies of the properties of radioligand dissociation can provide useful insights into mechanisms of receptor binding, especially changes in receptor affinity elicited by allosteric modifiers of the receptor system. In this setting, the term "allosteric" is an operational term to denote "other site," and does not imply a thermodynamically defined mechanism, such as the model for allostery developed by Monod, Wyman and Changeux (cf. chapter 1). If the allosteric agent causes a change in receptor affinity for the radioligand manifest in steady-state binding analysis that is due, at least in part, to changes in the rate of radioligand dissociation, then monitoring radioligand dissociation in the absence versus the presence of the putative allosteric modifier will permit direct detection of a modifier-dependent change in the radioligand dissociation rate.

The nature of the overall experimental design is shown schematically in figure 3-12. In these experiments, radioligand is added to the incubation and steady state binding is achieved. Just as for studies in figure 3-10, it is not necessary, but simply convenient, to achieve steady state binding before initiating the dissociation phase of these experiments. Specific binding is determined at the end of the association phase and dubbed as t_0 binding, or 100% of control binding at time = 0, relative to the dissociation phase. In this

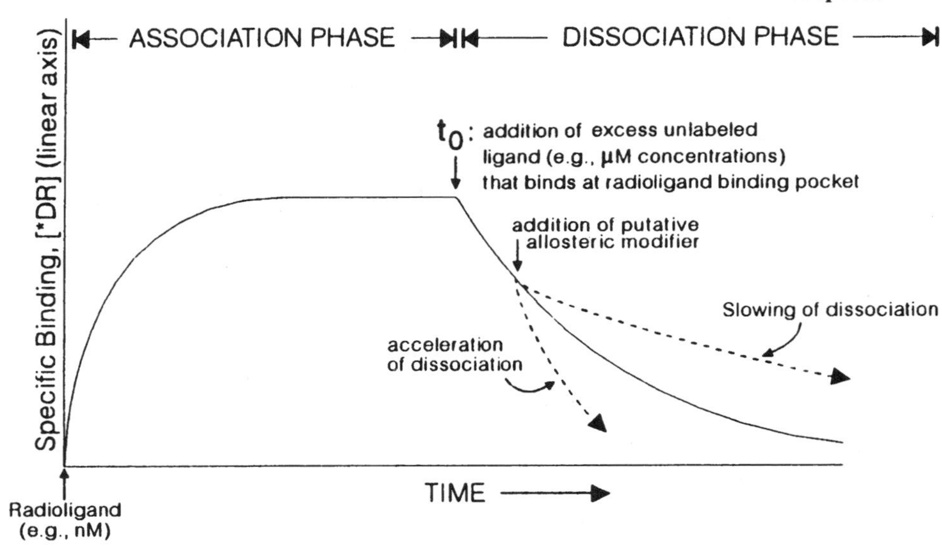

Figure 3-12. Changes in the rate of radioligand dissociation as a means to identify possible allosteric modulation of receptor-radioligand interactions. The experimental paradigm is described in detail in the text. Once it has been verified empirically that the concentration of excess unlabeled ligand added to initiate monitoring of the dissociation phase is sufficient to prevent both radioligand association and rebinding, once dissociated, then detection of a change in rate of radioligand dissociation upon addition of the allosteric modifier is evidence that this agent indeed alters receptor interaction with the radioligand by binding to a distinct or other (i.e., "allosteric") site not occupied by the radioligand. In non-purified systems, allosteric effects can be detected due to interactions with binding sites on the receptor molecule itself or due to interactions with binding sites on receptor-associated regulatory proteins propagated via conformational changes to the radioligand-binding receptor.

---

experimental paradigm, the dissociation phase is initiated by the addition of excess unlabeled ligand that binds at a site orthosteric to the radioligand. To test whether the concentration of unlabeled ligand is sufficient to fully occupy all radioligand binding sites and thus prevent ligand rebinding, a higher concentration of the same unlabeled agent can be added *or* a high concentration of another agent known to interact competitively at the radioligand binding site can be added. In either case, the lack of acceleration in the rate of radioligand dissociation is evidence that the concentration of excess unlabeled ligand first added to initiate detection of the dissociation phase is sufficient to prevent detection of rebinding the radioligand. At this point in the incubation, the orthosteric binding pocket of the receptor population under study is fully occupied, either by radioligand or by a competing, unlabeled ligand. Assuming that the putative allosteric modifier influences receptor affinity for radioligand at least in part by changes in the rate of radioligand dissociation, the introduction of this agent into the incubation would be predicted to accelerate dissociation, if overall receptor affinity for the radioligand is reduced by the allosteric modifier, or to slow the rate of dissociation, if overall receptor affinity for the radioligand is enhanced by the allosteric modifier. This experimental strategy has been applied to the characterization of multiple, independent allosteric binding sites for drugs interacting with voltage-gated Ca^{2+} channels (Garcia et al. [1986]) and to the allosteric sites for monovalent cations and amphipathic drugs, such as amiloride analogs, to G-protein-coupled receptors (Horstman et al. [1990]; Leppik [1998]). However, this strategy can also reveal conformationally propagated changes in receptor affinity due to binding to receptor-associated regulatory proteins. The most prominent example of this indirect type of "allosteric" modulation of receptor affinity via receptor-associated regulatory proteins is the ability of guanine nucleotide binding to the α subunits of heterotrimeric GTP-binding proteins to decrease affinity for agonist agents at G-protein-coupled receptors, manifested by guanine nucleotide-accelerated dissociation of radiolabeled agonists from receptors functionally coupled to G-proteins (cf. Neubig et al. [1985] as an example).

SUMMARY

This chapter has woven the mechanics of data acquisition with a description of the manner in which these data are analyzed to provide several independent lines of evidence that the radioligand binding observed is what one would expect for binding to the physiologically relevant receptor. The text has emphasized the assumptions made in deriving the algebraic descriptions that form the basis for data analysis by graphical or computer-assisted means because *only* when these assumptions are met can straightforward interpretations of the data be valid.

There are specified criteria of saturability, specificity, and appropriate kinetics that must be met to establish with confidence that a radioligand binding assay really monitors interactions at the physiologically relevant receptor. However, an essential additional criterion is that the binding data must be internally consistent to be credible. For example, if one obtains a linear Scatchard plot, one expects a Hill slope that equals 1. One also expects that transformations of kinetic data that permit the determination of the rate of radioligand association and dissociation are also linear. Similarly, the K_D value determined in equilibrium saturation binding studies should be comparable to the K_D value calculated for the same (albeit unlabeled) ligand from competition binding studies and to the K_D value calculated from kinetic studies. Data that are not internally consistent suggest that technical artifacts or inappropriate assumptions may be influencing the data analysis. The inappropriateness of the assumptions usually means that something more complicated than $D + R \rightleftharpoons DR$ is occurring. Importantly, inappropriate assumptions of a simple, reversible bimolecular reaction driven by mass action law may mask interesting biological complexities due to binding site diversity, multiple receptor affinity states or allosteric modifiers. Such interesting possibilities mean that further, and hopefully independent, experimental findings are needed to establish the basis for the apparent complexity of radioligand binding. Chapter 4 focuses in greater detail on biological situations where radioligand binding data would be expected to reflect a greater complexity and on approaches for discriminating among possible explanations for these complexities.

REFERENCES

General

Cheng, Y. and Prusoff, W.H. (1973) Relationship between the inhibition constant (K_I) and the concentration of an inhibitor that causes a 50% inhibition (I_{50}) of an enzymatic reaction. Biochem. Pharmacol. 22:3099-3108.

Chou, T.-C. (1974) Relationships between inhibition constants and fractional inhibition in enzyme-catalyzed reactions with different numbers of reactants, different reaction mechanisms, and different types and mechanisms of inhibition. Mol. Pharmacol. 10:235-247.

Christopoulos, A. and Kenakin, T. (2002) G Protein-Coupled Receptor Allosterism and Complexing. Pharmacol. Rev. 54:323-374.

Cornish-Bowden, A. and Koshland, D.E. (1975) Diagnostic uses of the Hill (Logit and Nernst) plots. J. Mol. Biol. 95:201-212.

Janin, J. (1973) The study of allosteric proteins. In *Progress in Biophysics and Molecular Biology*, J.A.V. Butler and D. Noble (eds.), pp. 77-119. New York: Pergamon Press.

Kenakin, T.P. (2001) Quantitation in Receptor Pharmacology. Rec. & Chan. 7:371-385.

Klotz, I.M. and Hunston, D.L. (1971) Properties of graphical representations of multiple classes of sites. Biochem. 10:3065-3069.

Koshland, D.E. (1970) The molecular basis for enzyme regulation. In *The Enzymes*, P.D. Boyer (ed.), chapter 7, pp. 341-396. New York: Academic Press.

Leppik, R.A., Lazareno, S., Mynett, A. and Birdsall, N.J.M. (1998) Characterization of the allosteric interactions between antagonists and amiloride analogues at the human α_{2A}-adrenergic receptor. Mol. Pharm. 53:916-925.

Scatchard, G. (1949) The attractions of proteins for small molecules and ions. Ann. N.Y. Acad. Sci. 51:660-672.

Whitehead, E. (1970) The regulation of enzyme activity and allosteric transition. Prog. Bioph. Mol. Biol. 21:321-397.

Weber, G. (1992) *Protein Interactions*, pp. 1-287. New York: Routledge, Chapman and Hall.

Wyman, J. (1948) Heme proteins. Adv. Protein Chem. 4:407-531 (especially pp. 436-443 for derivation of the Hill equation).

Wyman, J. and Gill, S.J. (1990) *Binding and Linkage*, pp. 1-308. Mill Valley, CA: University Science Books.

Yamamura, H.I., Enna, S.J. and Kuhar, M.J. (1985) *Neurotransmitter Receptor Binding*, 2nd ed. NY:Raven Press.

Specific

Adair, G.S. (1925) The hemoglobin system: The oxygen dissociation curve of hemoglobin. J. Biol. Chem. 63:529-545.

Aranyi, P. (1980) Kinetics of the hormone-receptor interaction. Competition experiments with slowly equilibrating ligands. Biochim. Biophys. Acta 628:220-227.

Bearer, C.F., Knapp, R.D., Kaumann, A.J., Swartz, T.L. and Birnbaumer, L. (1980) Iodohydroxybenzylpindolol: Preparation, purification, localization of its iodine to the indole ring, and characterization as a partial agonist. Mol. Pharmacol. 17:328-338.

Burgisser, E., Hancock, A.A., Lefkowitz, R.J. and DeLean, A. (1981) Anomalous equilibrium binding properties of high-affinity racemic radioligands. Mol. Pharmacol. 19:205-216.

Burgisser, E., Lefkowitz, R.J. and DeLean, A. (1981) Alternative explanation for the apparent "two step" binding kinetics of high-affinity racemic antagonist radioligands. Mol. Pharmacol. 19:509-512.

Chang, J.-J., Jacobs, S. and Cuatrecasas, P. (1975) Quantitative aspects of hormone-receptor interactions of high affinity. Effect of receptor concentration and measurement of dissociation constants of labeled and unlabeled hormones. Biochim. Biophys. Acta 406:294-303.

Colowick, S.P. and Womack, F.C. (1969) Binding of diffusible molecules by macromolecules: Rapid measurement by rate of dialysis. J. Biol. Chem 244:774-777.

DeLean, A. and Rodbard, D. (1979) Kinetics of cooperative binding. In *Receptors: A Comprehensive Treatise*, R.D. O'Brien (ed.), pp. 1443-1490. New York: Plenum Press.

DeLean, A., Hancock, A.A. and Lefkowitz, R.J. (1982) Validation and statistical analysis of the computer modeling method for quantitative analysis of radioligand binding data for mixtures of pharmacological subtypes. Mol. Pharmacol. 21:5-16.

DeMeyts, P. Bianco, A. and Roth, J. (1976) Site-site interactions among insulin receptors. Characterization of the negative cooperativity. J. Biol. Chem. 251:1877-1888.

Ehlert, F.J., Roeske, W.R. and Yamamura, H.I. (1981) Mathematical analysis of the kinetics of competitive inhibition in neurotransmitter receptor binding assays. Mol. Pharmacol. 19:367-371.

Garcia, M.L., King, V.F., Siegl, P.K.S., Reuben, J.P. and Kaczorowski, G.J. (1986) Binding of Ca^{2+} entry blockers to cardiac sarcolemmal membrane vesicles.: Characterization of diltiazem-binding sites and their interaction with dihydropyridine and aralkylamine receptors. J. Biol. Chem. 261:8146-8157.

Hill, A.V. (1910) The possible effects of the aggregation of the molecules of haemoglobin on its dissociation curves. J. Physiol. 40:iv-vii.

Hill, A.V. (1913) The combinations of haemoglobin with oxygen and with carbon monoxide. I. Biochem. J. 7:471-480.

Hollenberg, M.D. and Cuatrecasas, P. (1979) Distinction of receptor from non-receptor interaction in binding studies. In *The Receptors: A Comprehensive Treatise*, R.D. O'Brien (ed.). New York: Plenum Press.

Horstman, D.A., Brandon, S., Wilson, A.L., Guyer, C.A., Cragoe, Jr. E.J., and Limbird, L.E. (1990) An aspartate conserved among G-protein receptors confers allosteric regulation of α_2-adrenergic receptors by sodium. J. Biol. Chem. 265(35):21590-21595.

Insel, P.A., Mahan, L.C., Motulsky, H.J., Stoolman, L.M. and Koachman, A.M. (1983) Time-dependent decreases in binding affinity of agonists for β-adrenergic receptors of intact S49 lymphoma cells. A mechanism of desensitization. J. Biol. Chem. 258:13597-13605.

Jacobs, S., Chang, K.-H. and Cuatrecasas, P. (1975) Estimation of hormone receptor affinity by competitive displacement of labeled ligand: Effect of concentration of receptor and of labeled ligand. Biochem. Biophys. Res. Commun. 66:687-692.

Klotz, I.M. (1982) Numbers of receptor sites from Scatchard graphs: Facts and fantasies. Science 217:1247-1249.

Koshland, D.E., Nemethy, G. and Filmer, D. (1966) Comparison of experimental binding data and theoretical models in proteins containing subunits. Biochem. 5:365-385.

Langmuir, I.J. (1918) The adsorption of gases on plane surfaces of glass, mica and platinum. J. Amer. Chem. Soc. 40:1361-1372.

Linden, J. (1982) Calculating the dissociation constant of an unlabeled compound from the concentration required to displace radiolabel binding by 50%. J. Cycl. Nucl. Res. 8:163-172.

Motulsky, H.J. and Mahan, L.C. (1984) The kinetics of competitive radioligand binding predicted by the law of mass action. Mol. Pharmacol. 25:1-9.

Monod, J., Wyman, J. and Changeaux, J.-P. (1965) On the nature of allosteric transitions: A plausible model. J. Mol. Biol. 12:88-118.

Motulsky H and Christopoulos A (2003) *Fitting Models to Biological Data Using Linear and Nonlinear Regression.* GraphPad Software, Inc., San Diego, CA.

Munson, P.J. (1983) Experimental artifacts and the analysis of ligand binding data: Results of the computer simulation. J. Receptor Res. 3:249-259.

Munson, P.J. and Rodbard, D. (1983) Number of receptor sites from Scatchard and Klotz graphs: A constructive critique. Science 220:979-981.

Neubig, R., Gantzos, R.D. and Brasier, R.S. (1985) Agonist and antagonist binding to α_2-adrenergic receptors in purified membranes from human platelets. Mol. Pharm. 28:475-486.

Pittman, R.N. and Molinoff, P.B. (1980) Interactions of agonists and antagonists with β-adrenergic receptors on intact L6 muscle cells. J. Cyclic Nucl. Res. 6:421-435.

Rosenthal, H.E. (1967) A graphic method for the determination and presentation of binding parameters in complex systems. Anal. Biochem. 20:525-532.

Taylor, S.I. (1975) Binding of hormones to receptors. An alternative explanation of nonlinear Scatchard plots. Biochem. 14:2357-2361.

Wyman, J. (1967) Allosteric linkage. J. Amer. Chem. Soc. 89:2202-2218.

4. COMPLEX BINDING PHENOMENA

The methods for acquisition and initial analysis of radioligand binding phenomena were summarized in chapter 3. It was demonstrated that equations for linear transformations of binding data were derived assuming that a reversible bimolecular reaction driven by mass action occurred between ligand and receptor, $*D + R \rightleftharpoons *DR.$. Consequently, when data transformations such as the Scatchard plot are nonlinear, Hill coefficients (n_H) do not equal 1.0, or competition binding curves are not of normal steepness, additional complexities are suggested. Chapter 3 also provided guidelines for evaluating whether technical artifacts were responsible for departure of the data from those expected for a simple bimolecular reaction. Once technical artifacts have been excluded, complex binding phenomena suggest the existence of *biological* complexities that may provide insights into the molecular basis of receptor function.

This chapter will first summarize two general mathematical descriptions for complex binding phenomena, and indicate the assumptions inherent in each description. The use of non-linear regression analyses for obtaining binding parameters such as K_D and $B_{\max}$ values will then be described. A particular emphasis will be made regarding the appropriateness of the mathematical model inherent in computer programs that describe and quantitate the molecular phenomena being studied. The experimental and analytical approaches for differentiating discrete receptor subpopulations from interconvertible affinity states of a single receptor population then will be discussed. Finally, expansions of affinity state models to accommodate experimental findings in G protein-coupled receptor systems will be examined.

MATHEMATICAL DESCRIPTIONS OF COMPLEX BINDING PHENOMENA

All algebraic equations that have been used to describe complex binding phenomena have inherent assumptions. Understanding the various mathematical models proposed to describe the properties of complex radioligand binding data is important, as these models form the basis of computer-assisted non-linear regression analyses for evaluating these data. Most important for the investigator is being confident that the mathematical models, and their inherent assumptions, accurately reflect the biology of the system under study.

A description of the binding of ligand $*D$ to multiple sites on receptor R can be derived using a statistical approach. The statistical approach was inherent in the model introduced by Adair in 1925 to describe the binding of oxygen to hemoglobin. The model of Adair extended that of A. V. Hill by including all of the possible intermediates of the reaction between hemoglobin and oxygen, rather than assuming that only the empty and fully liganded forms of hemoglobin existed at equilibrium (Adair [1925]). (For a didactic elaboration of this model, see Newsholme and Start [1973]; for other theoretical development, see Janin [1973]; Koshland, Nemethy and Filmer [1966], and Teipel and Koshland [1969].)

The derivation below is a paraphrase of the Adair model, where the interaction being measured is the binding of radioligand $*D$ to a tetrameric receptor R, rather than the binding of oxygen to hemoglobin. If binding could be measured on each monomer in the absence of binding to sites on other monomers, the microscopic association constant for each site would be obtained. In a sense, this is an imaginary number, because binding at one site may influence binding at a second site. Since, in practice, the apparent association constant for each site varies according to the number of available binding sites on each polymer, microscopic and apparent association constants are *statistically* related. In the mathematical treatment to follow, the *probability* of binding at each site is emphasized.

Definitions:

K_{a_i} = microscopic or "intrinsic" association constant; identical for each monomer in the absence of interactions between sites

$K_{a_1}, K_{a_2}, K_{a_3}, K_{a_4}$ = "apparent" association constants

K_{a_i} is *statistically* related to the apparent binding constants by the number of empty binding sites available on the receptor molecule.

The statistical relationship between the microscopic and apparent association constants is based on calculable probabilities for association and dissociation of ligand from each site on the tetrameric receptor.
In the reaction of D with an empty tetramer:

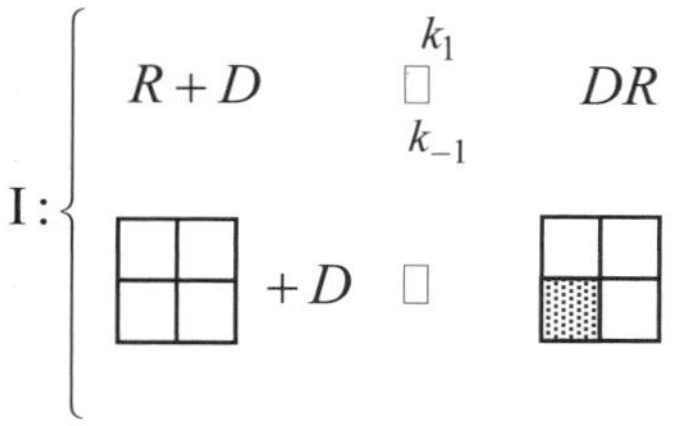

$$\text{I}: \quad R + D \underset{k_{-1}}{\overset{k_1}{\rightleftharpoons}} DR$$

Ligand binding sites on R:

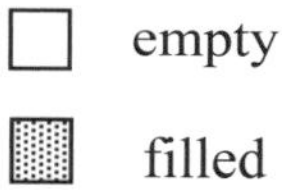

□ empty

■ filled

$$\text{Since } K_{a_i} = \frac{k_1}{k_{-1}} = \frac{k_2}{k_{-2}} = \frac{k_3}{k_{-3}} = \frac{k_4}{k_{-4}}$$

$$K_{a_1} = \frac{4k_1}{k_{-1}} = 4K_{a_i}$$

There are four sites available for binding (association) but only *one* from which D will dissociate. Therefore, K_{a_1} is four times greater than K_{a_i} (which corresponds to *one* site for both association and dissociation).

Similarly, in a stepwise fashion:

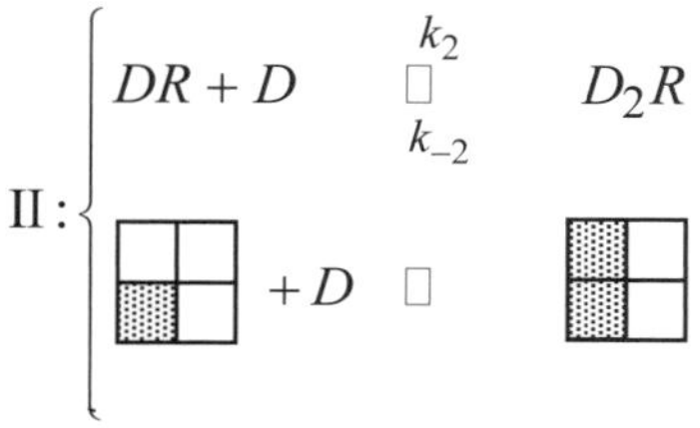

$$\text{II}: \quad DR + D \underset{k_{-2}}{\overset{k_2}{\rightleftharpoons}} D_2R$$

$$K_{a_2} = \frac{3\,k_2}{2\,k_{-2}} = \frac{3}{2} K_{a_i}$$

i.e., there are 3 sites to which D can associate, and 2 sites from which D can dissociate

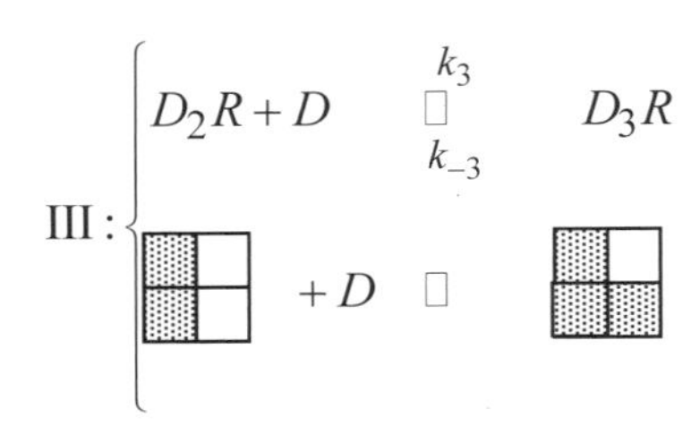

$$K_{a_3} = \frac{2\,k_3}{2\,k_{-3}} = \frac{2}{3} K_{a_i}$$

i.e., there are 2 sites to which D can associate, and 3 sites from which D can dissociate

IV : $D_3R + D \underset{k_{-4}}{\overset{k_4}{\rightleftharpoons}} D_4R$

$+D$

$$K_{a_4} = \frac{k_4}{4\,k_{-4}} = \frac{1}{4} K_{a_i}$$

i.e., there is one site to which D can associate, and 4 sites from which D can dissociate

Defining fractional saturation of the receptors (Y_s) as the total amount of ligand bound divided by the total number of *sites* available, one obtains:

$$Y_s = \frac{[DR] + 2\,[D_2R] + 3\,[D_3R] + 4\,[D_4R]}{4\,([R] + [DR] + [D_2R] + [D_3R] + [D_4R])} \tag{4.1}$$

The symbol Y_s (*sites* fractional occupancy) differs in meaning from that of Y used previously, because here Y_s is determined in terms of the number of *sites* available, whereas *molar* fractional occupancy (Y) was defined as moles of ligand bound per total *moles* of receptor molecules available. Since Adair's model arose as an attempt to describe the intermediate species that existed upon the binding of O_2 to hemoglobin (a tetrameric molecule) the number of sites in this derivation = $4[R]_{TOT}$.

The various forms of receptor-ligand complexes can be described in terms of K_A values $[R]$ and $[D]$, as shown below:

$$[DR] = K_{a_1}\,[R][D]$$
$$[D_2R] = K_{a_2}\,[DR][D] = K_{a_1} K_{a_2}\,[R][D]^2$$
$$[D_3R] = K_{a_3}\,[D_2R][D] = K_{a_1} K_{a_2} K_{a_3}\,[R][D]^3$$
$$[D_4R] = K_{a_4}\,[D_3R][D] = K_{a_1} K_{a_2} K_{a_3} K_{a_4}\,[R][D]^4$$

Substituting for the species identified in equation 4.1 in terms of K_A, $[R]$ and $[D]$:

$$Y_s = \frac{K_{a_1}[R][D] + 2K_{a_1}K_{a_2}[R][D]^2 + 3K_{a_1}K_{a_2}K_{a_3}[R][D]^3 + 4K_{a_1}K_{a_2}K_{a_3}K_{a_4}[R][D]^4}{4\,([R] + K_{a_1}[R][D] + K_{a_1}K_{a_2}[R][D]^2 + K_{a_1}K_{a_2}K_{a_3}[R][D]^3 + K_{a_1}K_{a_2}K_{a_3}K_{a_4}[R][D]^4)}$$

Since $[R]$ is in all the terms in the numerator and denominator, one can simplify to:

$$Y_s = \frac{K_{a_1}[D] + 2K_{a_1}K_{a_2}[D]^2 + 3K_{a_1}K_{a_2}K_{a_3}[D]^3 + 4K_{a_1}K_{a_2}K_{a_3}K_{a_4}[D]^4}{4\,(1 + K_{a_1}[D] + K_{a_1}K_{a_2}[D]^2 + K_{a_1}K_{a_2}K_{a_3}[D]^3 + K_{a_1}K_{a_2}K_{a_3}K_{a_4}[D]^4)} \quad (4.2)$$

The above polynomial expression often is referred to as **Adair's equation** and was derived without any assumptions concerning the independence of binding at different ligand-combining sites. The derivation similarly did not dictate any necessary relationship between the intrinsic affinities (K_{a_i} values) at each site. It can be seen that in situations where there are *no* interactions among binding sites, such that each binding interaction occurs independently of all other ligand-protein interactions, $K_{a_1} \cdots K_{a_4}$ are related to K_{a_i} by constant factors:

$$K_{a_1} = 4K_{a_i}$$

$$K_{a_1}K_{a_2} = 4 \cdot 3/2 \cdot K_{a_i}K_{a_i} = 6K_{a_i}^{\ 2}$$

$$K_{a_1}K_{a_2}K_{a_3} = 4 \cdot 3/2 \cdot 2/3 \cdot K_{a_1} \cdot K_{a_i} \cdot K_{a_i} = 4K_{a_i}^{\ 3}$$

$$K_{a_1}K_{a_2}K_{a_3}K_{a_4} = 4 \cdot 3/2 \cdot 2/3 \cdot 1/4 \cdot K_{a_i} \cdot K_{a_i} \cdot K_{a_i} \cdot K_{a_i} \cdot = K_{a_i}^{\ 4}$$

and substituting these relationships into equation 4.2 yields:

$$Y_s = \frac{4K_{a_i}[D] + 12K_{a_i}^{\ 2}[D]^2 + 12K_{a_i}^{\ 3}[D]^3 + 4K_{a_i}^{\ 4}[D]^4}{4\,(1 + 4K_{a_i}[D] + 6K_{a_i}^{\ 2}[D]^2 + 4K_{a_i}^{\ 3}[D]^3 + K_{a_i}^{\ 4}[D]^4)} \quad (4.3)$$

$$Y_s = \frac{4K_{a_i}[D](1 + 3K_{a_i}[D] + 3K_{a_i}^{\ 2}[D]^2 + K_{a_i}^{\ 3}[D]^3)}{4\,(1 + 4K_{a_i}[D] + 6K_{a_i}^{\ 2}[D]^2 + 4K_{a_i}^{\ 3}[D]^3 + K_{a_i}^{\ 4}[D]^4)} \quad (4.4)$$

Note that

$$(1 + K_{a_i}[D])^3 = (1 + 3K_{a_i}[D] + 3K_{a_i}^{\ 2}[D]^2 + K_{a_i}^{\ 3}[D]^3)$$

and that

$$(1 + K_{a_i}[D])\,(1 + K_{a_i}[D])^3 = (1 + 4K_{a_i}[D] + 6K_{a_i}{}^2[D]^2 + 4K_{a_i}{}^3[D]^3) + K_{a_i}{}^4[D]^4)$$

so that

$$Y_S = K_{a_i}[D]\frac{(1+K_{a_i}[D])^3}{(1+K_{a_i}[D])^4} = \frac{K_{a_i}[D]}{1+K_{a_i}[D]} \tag{4.5}$$

The relationship in equation 4.5 indicates that when a ligand interacts with binding sites in an entirely statistical fashion (i.e., the binding sites are identical and there are no interactions among the sites that modify binding properties at any site), the algebraic description of these interactions is equivalent to the Langmuir binding isotherm (or the Michaelis-Menten equation when $k_3 = 0$) for a monomeric protein, and the data describe a hyperbolic curve. In this situation K_{a_i} is equivalent to the equilibrium association constant K_A. Even in the *absence* of interactions among the binding sites, however, the statistical relationship linking the step-wise association constants is the following:

For a tetramer:

$$\frac{K_{a_1}}{K_{a_2}} = \frac{4}{3/2} = \frac{8}{3}$$

$$\frac{K_{a_2}}{K_{a_3}} = \frac{3/2}{2/3} = \frac{9}{4}$$

$$\frac{K_{a_3}}{K_{a_4}} = \frac{2/3}{1/4} = \frac{8}{3}$$

K_{a_2} is therefore 3/8 of K_{a_1}, K_{a_3}= 4/9 of K_{a_2}, and K_{a_4} = 3/8 of K_{a_3}.

In the presence of cooperativity, these ratios will be modified. For positive cooperativity, where binding of D to DR is facilitated by binding of D to the first site on R, K_{a_2} will be greater than 3/8 of the value for K_{a_1}, etc. The reverse is true for negative cooperativity.

The same polynomial expression found in equation 4.2 can be derived in another way, by simple algebraic substitution into equations defining the equilibrium association constant. For example, Klotz (1946) demonstrated that when a ligand interacts with multiple, independent ligand-combining sites on a single protein, the interactions can be described as follows:

$$R + D \overset{K_{A_1}}{\rightleftharpoons} DR \quad \text{and} \quad K_{A_1} = [DR]/[R][D]$$
$$DR + D \overset{K_{A_2}}{\rightleftharpoons} D_2R \quad \text{and} \quad K_{A_2} = [D_2R]/[DR][D]$$
$$D_2R + D \overset{K_{A_3}}{\rightleftharpoons} D_3R \quad \text{and} \quad K_{A_3} = [D_3R]/[D_2R][D]$$
$$D_3R + D \overset{K_{A_4}}{\rightleftharpoons} D_4R \quad \text{and} \quad K_{A_4} = [D_4R]/[D_3R][D]$$

If $Y = \dfrac{\text{moles of ligand bound}}{\text{total moles of receptor available}}$

$$Y = \frac{[DR]+2[D_2R]+3[D_3R]+4[D_4R]}{[R]+[DR]+[D_2R]+[D_3R]+[D_4R]} \tag{4.6}$$

If this expression for fractional occupancy is expressed in terms of K_A values, $[D]$, and $[R]$, then, as shown above,

$$[\text{DR}] = K_{A_1}\,[R][D]$$
$$[D_2R] = K_{A_2}\,[DR][D] = K_{A_1}\,K_{A_2}\,[R][D]^2$$
$$[D_3R] = K_{A_3}\,[D_2R][D] = K_{A_1}\,K_{A_2}\,K_{A_3}\,[R][D]^3$$
$$[D_4R] = K_{A_4}\,[D_3R][D] = K_{A_1}\,K_{A_2}\,K_{A_3}\,K_{A_4}\,[R][D]^4$$

By substituting the above expressions into equation 4.6 and dividing through by $[R]$,

$$Y = \frac{K_{A_1}[D]+2K_{A_1}K_{A_2}[D]^2+3K_{A_1}K_{A_2}K_{A_3}[D]^3+4K_{A_1}K_{A_2}K_{A_3}K_{A_4}[D]^4}{1+K_{A_1}[D]+K_{A_1}K_{A_2}[D]^2+K_{A_1}K_{A_2}K_{A_3}[D]^3+K_{A_1}K_{A_2}K_{A_3}K_{A_4}[D]^4} \tag{4.7}$$

which can be restated by the general expression:

$$Y = \frac{K_{A_1}[D]+2K_{A_1}K_{A_2}[D]^2+\ldots}{1+K_{A_1}[D]+K_{A_1}K_{A_2}[D]^2\ldots} \tag{4.8}$$

Like the Adair equation, the polynomial expression in equation 4.8 is always valid for correlating binding data, regardless of the molecular model. Hence, equation 4.8 is a valid mathematical model for describing binding when all sites are identical, when discrete and independent populations of binding sites possessing different affinities for ligand exist, or when there is negative cooperativity or positive cooperativity or both. As demonstrated by Klotz (1983), by dividing both sides of equation 4.8 by $[D]$ (the concentration of free ligand), the first stoichiometric binding constant (K_{A_1}) can be evaluated graphically by plotting $Y/[D]$ versus $[D]$ and extrapolating $[D]$ to zero. Unfortunately, there are no short-cut methods for evaluating succeeding association constants K_{A_2}, K_{A_3} and K_{A_4}.

When multiple populations of binding sites possessing different affinities for ligand exist (e.g., R_1, R_2, R_3) and bind ligand independently of one another, then the binding observed can be appropriately described as a **sum of hyperbolas**, with each hyperbolic equation representing the quantity of binding observed at each site, where n_1 = number of binding sites for receptor population R_1, n_2 = number of binding sites for receptor population R_2, etc. If each receptor has one ligand-combining site, then $n_1 = [R_1]$ and $n_2 = [R_2]$, etc.:

$$[DR] = \frac{n_1[D]}{K_{D_1} + [D]} + \frac{n_2[D]}{K_{D_2} + [D]} + \frac{n_3[D]}{K_{D_3} + [D]} \qquad (4.9)$$

or, in terms of equilibrium association constant K_A:

$$[DR] = \frac{n_1 K_{A_1}[D]}{1 + K_{A_1}[D]} + \frac{n_2 K_{A_2}[D]}{1 + K_{A_2}[D]} + \frac{n_3 K_{A_3}[D]}{1 + K_{A_3}[D]} \qquad (4.10)$$

There are two limitations, however, to treating observed binding data as the sum of multiple hyperbolic binding functions. First, the investigator must have independent data to confirm that the sites indeed behave as independent receptor populations, and that each of these receptor populations binds ligand via mass action law. Second, it can be shown that a large difference in K_A values (10^3-10^4) is necessary to completely resolve data for two populations of sites from one another (Klotz [1983], Steinhardt and Reynolds [1969]), so that the binding does appear as independent hyperbolic functions.

Actually, it is worth noting that equation 4.2 (and thus equation 4.8) can be converted by algebraic manipulation to a form resembling the sum of hyperbolas shown in equation 4.9 (Klotz and Hunston [1975]).

$$Y = \frac{k_\alpha[D]}{1 + k_\alpha[D]} + \frac{k_\beta[D]}{1 + k_\beta[D]} + \frac{k_\gamma[D]}{1 + k_\gamma[D]} \qquad (4.11)$$

However, in this case, the parameters k_α, k_β and k_γ are constants but are *not* site-binding constants, except in the special case where multiple, discrete, and entirely independent ligand-combining sites exist. Otherwise, these k_α, k_β and k_γ values, to paraphrase Klotz and Hunston (1984), are parameters for "ghost sites," i.e., imaginary, nonexistent sites that can be assigned binding constants which, when inserted into equation 4.11, can reproduce the observed binding data but do not necessarily reflect thermodynamic constants for *real* protein-ligand interactions. Stated another way, in the absence of independent lines of experimental data documenting that two or more classes of sites with fixed (but different) affinities exist, the k_α, k_β and k_γ parameters obtained by analyzing data in this manner are purely empirical values that have no precise thermodynamic meaning. It is imperative that the investigator seek independent data documenting the existence of independent receptor populations before assigning ghost site terms to presumptive physical realities. Strategies that can be employed to test the mathematical model to affirm whether discrete receptor subpopulations exist are discussed later in this chapter.

NON-LINEAR REGRESSION ANALYSIS OF COMPLEX BINDING PHENOMENA

As indicated in chapter 3, there are many possible molecular mechanisms that can account for complex radioligand binding phenomena. These complex phenomena deviate from a simple bimolecular reaction driven solely by mass action in two general ways. One way for deviation to occur is due to positively cooperative binding, which results when the affinity of the receptor population increases with increasing fractional occupancy of the receptors. This is manifested by concave downward Scatchard plots, Hill plots with n_H values > 1.0, or competition binding profiles with slope factors > 1.0. Apparent positively cooperative binding phenomena are not commonly observed in radioligand binding studies. The second, and more frequently observed, deviation from binding to a single receptor with a single unchanging affinity are binding data demonstrating concave upward Scatchard plots, Hill plots with n_H values < 1.0, and so-called "shallow" competition binding profiles with slope factors < 1.0. These latter observations can reflect the occurrence of numerous molecular phenomena, including (1) negative cooperativity among binding sites, such that the overall affinity of the receptor population decreases as fractional occupancy increases; (2) multiple independent populations of receptors or binding sites with discrete and unchanging affinities for ligands; or (3) multiple affinity states of the receptor for ligand, such as those resulting from a two-step reaction involving formation of a ternary complex (i.e., $D + R \rightleftharpoons DR + X \rightleftharpoons DRX$).

Unfortunately, equilibrium binding data cannot distinguish among these latter possibilities. Kinetic strategies, as discussed in chapter 3, can be informative, but multiple independent lines of biochemical evidence should be sought that can discriminate among the possible explanations for heterogeneous ligand-receptor interactions to obtain a plausible molecular model that accounts for the observed binding data. An appropriate mathematical model consistent with the postulated molecular model then can be derived or applied. With an appropriate mathematical model, the investigator can either program a computer or use an available computer program to analyze binding data and obtain useful parameters describing the ligand-receptor interactions, such as K_D and $[R]_{TOT}$ values.

A variety of computer programs are available for the analysis of radioligand binding data. Like the mathematical descriptions of complex binding phenomena on which they are based, commonly used computer programs for analyzing radioligand binding data fall into two general categories: nonrestrictive (analogous to the empirical Hill equation, equation 3.22) and restrictive (analogous to equation 4.9 or 4.10).

One example of a nonrestrictive mathematical model used as a basis for computer-assisted analysis of binding data is a four-parameter logistic equation which, as indicated above, is mathematically analogous to the Hill equation (see DeLean et al. [1978]). In this model:

$$y = \frac{a-d}{1+(X/c)^b} + d \tag{4.12}$$

where y = response

X = the dose of agonist

a = the response when $X = 0$

d = the response for an "infinite" dose

c = EC_{50}, the dose resulting in a response halfway between a and d

b = a "slope factor" that describes the steepness of a curve. This factor corresponds to the slope of a logit-log plot when X is portrayed in terms of natural logarithms.

Programs based on this model are especially useful for the analysis of families of curves obtained when competition binding studies are performed with a variety of unlabeled competitors. In this situation, the above parameters can be specified as follows:

Y = concentration of radioligand bound

X = concentration of competitor bound

c = EC_{50} of competitor

b = steepness factor or slope factor
a = extrapolated upper limits for Y (analogous to "total binding")
d = extrapolated lower limits for Y (analogous to "nonspecific binding")

Some practical advantages of computer-assisted analysis of radioligand binding data are immediately apparent when considering equation 4.12. First, the fourth parameter (d) is a determination of nonspecific binding. Because computer data are weighted as a reciprocal of their variance, nonspecific binding as well as total binding can be extrapolated by relying most heavily on those data points that are obtained with greatest experimental accuracy. Consequently, nonspecific binding can be determined based on the characteristics of all of the data rather than as a result of a somewhat arbitrary definition (see the discussion concerning assessment of nonspecific binding in chapter 3). In equation 4.12, the slope factor (b) permits a quantitative expression of the curve shape. A slope factor of 1.0 is consistent with ligand-receptor interactions occurring via a reversible bimolecular interaction that obeys mass action law, whereas slope factors > 1.0 may indicate positive cooperativity and those < 1.0 may indicate negative cooperativity, receptor heterogeneity, or multiple receptor affinity states. (It should be remembered that for competition data, the sign of the slope will be negative in a manner analogous to indirect Hill plots; cf. equation 3.26 or logit-log plots.) Although the slope factor has the same mathematical form as the Hill coefficient (n_H), it should not be interpreted in the same thermodynamic terms, except under special circumstances. For example, the value of X normally used in these computations is the total concentration of ligand added to an incubation. In contrast, the Hill analysis would require the concentration of free radioligand to be determined and employed for computation. The ability to use the concentration of ligand added to the incubation using the four-parameter logistic equation is one major advantage over the Hill method, since errors introduced into the parameter estimates due to poor estimation of the concentration of free radioligand are eliminated. Finally, computer programs based on the four-parameter logistic equation allow the investigator to consider each competition binding curve individually or to analyze all of the curves simultaneously. In the latter case, the investigator forces the curves to share certain parameters-for example, slope factors-and can thereby determine, using a statistical analysis of the "goodness of fit" of data when curves are constrained in this way, whether two or more ligands interact with the receptor to the same degree of complexity. "Constrained" curve fitting not only may provide more information regarding ligand-receptor interactions, but also may be necessary in some cases to permit the curve-fitting routine to provide appropriate parameter estimates, since data in a particular part of the curve may be absent for some, but not all, experiments performed using an identical protocol (DeLean et al. [1978]; Motulsky and Christopoulos [2003]).

To reiterate, the two parameters describing receptor-ligand interactions that can be obtained from the four-parameter logistic equation based on the empirical Hill equation are (1) the slope factor describing the shape of the competition curve and (2) an EC_{50} value for the midpoint of the curve. If the slope factor equals 1.0 or -1.0, the data are consistent with the conclusion that the interaction between ligand and receptor can be described by a reversible bimolecular reaction obeying mass action law. In this case, it is valid to calculate a K_D value for receptor-ligand interactions from the EC_{50} value using an approximation such as the Cheng and Prusoff/Chou equation (see equation 3.27). Analysis of radioligand binding data using the four-parameter logistic equation, however, cannot resolve complex binding phenomena further, for example, into two or more populations of binding sites or affinity states. Consequently, several computer modeling programs have been developed based on equation 4.9 (or 4.10) to permit calculation of additional descriptive parameters for complex radioligand binding data.

The mathematical model on which many programs for analysis of radioligand data are based is analogous to the "sum of hyperbolas" description for complex binding phenomena given in equation 4.10 (Motulsky and Christopoulos [2003]; Munson and Rodbard [1980]; Munson [1983]; Rodbard [1973]). The inherent model is the general "$N \times M$" model for N ligands binding to M classes of receptor sites (Feldman [1972]). This general relationship can be described in more specific terms for the two types of data usually submitted to computer analysis: **saturation binding data**, where the receptor population(s) is occupied by increasing concentrations of a radiolabeled ligand, and **competition binding data**, where the receptor is confronted with both a radiolabeled ligand and a competing, unlabeled ligand. The algebraic descriptions that follow are from Munson (1983). Please note that the K value represents the equilibrium *association* constant, in units of M^{-1}.

1. For a single ligand binding to a single class of binding sites:

$$B = [KR/(1 + KF) + N]F \qquad (4.13)$$
$$T = B + F$$

where T = concentration of total ligand added
B = concentration of bound ligand
F = concentration of free ligand
R = receptor density
N = ratio of nonspecifically bound to free ligand
K = equilibrium association constant, in units of M^{-1}

2. For a single ligand binding to two independent classes of receptors:

$$B = [K_1R_1/(1 + K_1F) + K_2R_2/(1 + K_2F) + N]F \tag{4.14}$$

The extension to several independent classes would involve addition of the appropriate number of hyperbolic functions: $K_nR_n/(1 + K_nF)$.

3. For two ligands binding to a single class of receptors (as occurs in a competition binding study), the mathematical model becomes:

$$B_1 = [K_1R/(1 + K_1F_1 + K_2F_2) + N_1]F_1 \tag{4.15}$$

$$B_2 = [K_2R/(1 + K_1F_1 + K_2F_2) + N_2]F_2 \tag{4.16}$$

$$T_1 = B_1 + F_1$$

$$T_2 = B_2 + F_2$$

where the subscript on T, B, K and F refers to ligand L_1 or L_2. The value of K_1 for radioligand L_1 is determined in independent experiments by analysis of saturation binding data.

The expressions for B_1 and B_2 in equations 4.15 and 4.16 differ from those in equations 4.13 and 4.14 because the amount of binding of one ligand to the receptor populations is necessarily influenced by the fractional occupancy of the receptor population attained by the other ligand. It can be shown that if there is one receptor population (or one set of binding sites) but two ligands competing for this set, then fractional occupancy with L_1 can be expressed as:

$$Y_1 = \frac{K_1[F_1]}{1 + K_1[F_1]}(1 - Y_2) \tag{4.17}$$

and fractional occupancy with L_2 can be expressed as:

$$Y_2 = \frac{K_2[F_2]}{1 + K_2[F_2]}(1 - Y_1) \tag{4.18}$$

Thus, equations 4.15 and 4.16 take into account that part of the receptor population will be filled by each ligand. (This derivation assumes that each receptor R has only one ligand combining site ($n = 1$), such that the total number of binding sites $n[R] \equiv [R]$.)

Algebraic combination of the above equations leads to:

$$Y_1 = \frac{K_1F_1}{1 + K_1F_1 + K_2F_2} \tag{4.19}$$

$$Y_2 = \frac{K_2F_2}{1+K_1F_1+K_2F_2} \tag{4.20}$$

and since $Y = \dfrac{B}{[R]_{TOT}}$, then $B_1 = Y_1R_1$ and $B_2 = Y_2R_2$ for substitution into equations 4.15 and 4.16.

4. **For two ligands binding to two classes of independent populations of receptors,** a double subscript is used in the LIGAND program to describe the affinity constant K:

$$B_1 = [K_{11}R_1/(1 + K_{11}F_1 + K_{21}F_2) + K_{12}R_2/(1 + K_{12}F_1 + K_{22}F_2) + N_1]F_1 \tag{4.21}$$

$$B_2 = [K_{21}R_1/(1 + K_{11}F_1 + K_{21}F_2) + K_{22}R_2/(1 + K_{12}F_1 + K_{22}F_2) + N_1]F_2 \tag{4.22}$$

This mathematical model then can be extended in an analogous manner to any number of ligands and any number of classes of sites.

Computer-assisted analyses based on these or similar equations typically introduce a correction factor that adjusts for varying receptor concentrations between experiments, permitting the simultaneous analysis and comparison of data obtained from several experiments. This correction factor (C) simply adjusts the binding of the second experiment (C_2) relative to the first. The mathematical description for comparing binding in two separate experiments for a one-ligand, one-binding site model becomes:

$$B_1 = [KR/(1 + KF_1) + N]F_1 \tag{4.23}$$

$$B_2 = [KR/(1 + KF_2) + N]F_2C_2 \tag{4.24}$$

$$T_1 = B_1 + F_1$$

$$T_2 = B_2 + F_2$$

where subscripts *1* and *2* refer to conditions in the first and second experiments, respectively. When only one experiment is performed, a value for C must nonetheless be assigned, and C_1 is set to equal 1. In addition, when specific and nonspecific binding do not vary proportionately between experiments, separate correction factors may be introduced for specific and nonspecific binding.

As emphasized earlier, computer-assisted non-linear regression analyses can weight the data based on the reciprocal of their variance, so the analysis is more significantly influenced by the most reliable data. Second, the computation is done using the concentration of ligand added to the incubation, which can be determined precisely. Consequently, all measurement error is confined to a single variable; the concentration of bound radioligand (B or $[^*DR]$). Third, nonspecific binding is not arbitrarily defined by the investigator, but estimated from the whole of the data. Fourth, the curve-fitting program provides a variety of statistical methods for evaluating the goodness of fit for a given model, e.g., a one-site versus a two-site model, and can therefore provide an objective assessment of the complexity of ligand-receptor interactions in light of the reliability of the raw data provided. The parameters for affinity constants and receptor densities also are provided with their standard errors, permitting an assessment of confidence limits for the parameters obtained. Finally, introduction of a correction factor (C) allows curves from several experiments to be considered simultaneously, which improves the statistical reliability of the data analysis and, hence, the validity of the results.

The following summarizes a potential strategy for analyzing competition binding data. First, data would be analyzed by the four-parameter logistic equation to provide an estimate of the curve's slope factor. Alternatively, a slope factor could be determined by fitting competitive binding data to equation 4.25 (Limbird and Motulsky [1998]):

$$\frac{(\text{Total} - \text{nonspecific})}{1+10^{\left(\log \text{IC}_{50} - \log[D]\right)\bullet \text{slope factor}}} \qquad (4.25)$$

Obtaining a slope factor of 1.0 would indicate that the data are consistent with a simple bimolecular interaction-one ligand interacting reversibly with one receptor population that possesses an unchanging affinity for ligand (the slope factor is negative because the curve goes downhill). The same data would be expected to be "best fit" by a one-site model, and a K_D as well as a B_{max} value could be estimated for this single receptor population. If a slope factor of < 1.0 were obtained from analysis of the data by the four-parameter logistic equation, then the EC_{50} obtained from this analysis would not correspond to the equilibrium binding constant, but would be an empirical value describing the midpoint position of the binding isotherm. Non-linear regression analysis of the data would be expected to demonstrate a better "fit" using a two-site model than a one-site model. In actuality, a two-site model almost always fits the data better than a one-site model, just as a three-site model fits even better, and so on. This is because as more variables (sites) are added to the equation, the curve becomes more "flexible" and aligns better with the experimental data points. Thus, it is essential to compare the improvement of "fit" of the data to a two-site, rather than a one-site, model using statistical analyses.[1] When the two models being compared are "nested," i.e. one model is a simpler case of the other, then testing which model (simpler or more complex) to accept is accomplished using the "*F* test," or by calculating an *F* ratio. If the models are not nested, the Akaike's Information Criterion Method, based on information theory, should be used (Motulsky and Christopoulos [2003]). An *F* ratio quantifies the relationship between the relative increase in the sum-of-squares (SS) and the relative increase in the degrees of freedom (DF).

$$F = \frac{\dfrac{(SS_1 - SS_2)}{SS_2}}{\dfrac{(DF_1 - DF_2)}{DF_2}} \text{ or } \frac{DF_2(SS_1 - SS_2)}{SS_2(DF_1 - DF_2)} \quad (4.25)$$

Where SS_1 and SS_2 are the sum-of-squares for one versus two site fits respectively and DF_1 and DF_2 are the degrees of freedom. If a one-site model is correct, the *F* ratio will be ~ 1.0. If *F* is significantly less than 1.0, then either a two-site model is correct, or the one-site model is correct, but by chance random scatter supports the two-site model. A *P* value can be calculated from the *F* ratio and the two degrees of freedom values. The *P* value provides an estimate of how rarely, or not, this coincidence would occur. The *P* value addresses the question of what is the chance that the data fit the two-site model so much better than could be obtained randomly. If the *P* value is small (e.g. $P \leq 0.05$), then it is reasonable to conclude that a two-site model, rather than a one-site model, is a significant improvement in the description of the data. The interpretation of these findings, even in light of statistical analyses, should also be influenced by common sense. Thus, a two-site fit might be disregarded if the second site has only a very small fraction of receptors, the K_D value for one of the sites is outside the range of the raw data, or the best fit values for the bottom and top plateaus of the competition binding data are far from the values actually observed in the experiment.

If statistical analysis is consistent with acceptance of a two site model, non-linear regression analysis can then provide parameter estimates for the *K* and *R* values at each of these "sites." If the assumptions of the mass action

[1] Typically, 15–18 data points are required on a competition curve to resolve, in a statistically significant fashion, a one-site from a two-site fit to the data. Birdsall and associates (1980) used over 50 concentrations of competitor to define three classes of muscarinic receptor in the medulla/pons. However, it is probable that more than three independent receptor populations, i.e., subtypes, cannot be defined in a given tissue, and three may be considered the upper limit. In practice, identifying two sites with confidence is likely the practical upper limit for subtype analysis using radioligand binding studies.

model inherent in the non-linear regression analysis are met by the biological system (namely, that receptor sites bind ligand independently of one another with an affinity that remains unchanged with increasing occupancy), then the binding parameters obtained will reflect the K_A (or $1/K_D$) and receptor density (B_{max}) values for each receptor population. As mentioned earlier, equilibrium binding data cannot ascertain whether independent populations of binding sites exist or whether the complex binding phenomena result from negative cooperativity or interconvertible affinity states of the receptor. It requires independent lines of biochemical evidence to resolve these issues. When the existence of independent receptor populations has not been unequivocally documented, the parameter estimates obtained using a mathematical model such as the "sum of hyperbolas" model may not have thermodynamic significance, and may simply be empirical descriptors analogous to the k_α, k_β, and k_γ parameters defined in equation 4.11. Nonetheless, these empirical parameters may be useful in comparing the nature of the biological system under differing experimental conditions.

Despite the unquestionable value of using computer-assisted non-linear regression analysis to evaluate radioligand binding data, it is unwise to disregard graphical methods for presenting and considering raw data, as discussed earlier in chapter 3. Graphical presentations are easier to interpret or understand intuitively than numerical parameters produced by computer-assisted analysis. Any data modeling also should be accompanied by a graphical output, so that the investigator can inspect whether the "best fit" obtained by computer analysis of the raw data generates a computer-drawn line that sensibly describes the trend in the binding data obtained.

INDEPENDENT RECEPTOR SUBTYPES

It is not uncommon to discover that a ligand interacts with a number of physically and functionally independent receptor populations. In some cases, this is a manifestation of ligand nonselectivity; for example, the ergot alkaloid dihydroergocryptine is an α-adrenergic antagonist in peripheral tissues but behaves as a dopamine agonist in the pituitary. Endogenous ligands also interact with multiple receptor populations. For example, insulin alters metabolic processes such as glucose transport via a receptor specific for insulin. However, at higher concentrations insulin also can interact with-and presumably modulate-cell function via distinct populations of receptors for insulin-like growth factors. Epinephrine is another example of an endogenous agent that interacts with several receptor populations, including multiple subtypes of α and β-adrenergic receptors and, at higher concentrations, dopamine receptors.

For most neurotransmitters, chemokines and autocrine agents, the existence of receptor subpopulations, or subtypes, is more often the rule than the exception. These subtypes are distinguished by the differing orders of potency of agonists and antagonists at these receptors, and by being encoded by distinct genes, splice variants, or edited versions of distinct genes. A database (http://www.gpcr.org) summarizes the known molecularly characterized G protein-coupled receptors (GPCRs).

There are three general approaches to identifying and quantifying receptor in a given target tissue using radioligand binding techniques. Two of these approaches rely on the existence of a ligand that is reasonably specific for one of the two subtypes such that saturating or near-saturating occupancy of one receptor subpopulation occurs without any detectable occupancy of the second subpopulation (see Lavin et al. [1981] as an example). Typically, such a highly selective ligand is not available. Consequently, a third approach has been developed that permits successful identification and characterization of receptor subtypes when ligands of only moderate selectivity for one of the two receptor subpopulations are available.

If a *radio*ligand *specific* for one of two putative receptor subpopulations is available, this radioligand can then be used to ascertain the existence of a particular receptor subtype in a target tissue of interest. If identified, the subpopulation can be characterized in terms of its affinity and density by a straightforward analysis of radioligand binding, as outlined in chapter 3.

If a specific ligand is available, but is *not* radiolabeled, this agent can be exploited in the following way: a saturation binding analysis of a radiolabeled agent that can interact with all subtypes under study is performed in the absence and presence of a concentration of the unlabeled, subtype-specific agent that should occupy all receptors of one subtype. The density of both subtypes can be determined by "subtraction" using least-squares regression: (1) total specific radioligand binding = binding to $R_1 + R_2$; (2) saturation binding in the presence of an unlabeled agent that presumably saturates one of the two receptor subtypes, e.g. R_2, permits an assessment of the affinity and density of the R_1 population (see figure 4-1A). The difference between the binding reflecting $R_1 + R_2$ and that reflecting R_1 represents binding contributed by the receptor population designated as R_2. This approach again relies on the availability of an unlabeled ligand with considerable specificity for one particular receptor subtype so that it can be used to selectively mask that particular subtype from occupancy by the radioligand.

A third, more general approach to characterizing receptor subtypes is to evaluate the ability of unlabeled agents that demonstrate some, but not absolute, subtype selectivity to compete for binding of a radiolabeled agent that can interact with both receptor subtypes. As shown schematically in figure 4-1B, competition binding studies are performed, and linear regression analyses based on equation 4.10 are performed, resulting in the estimation of

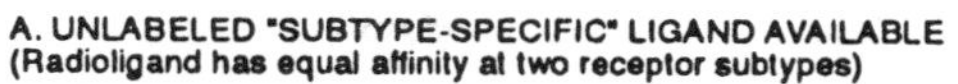

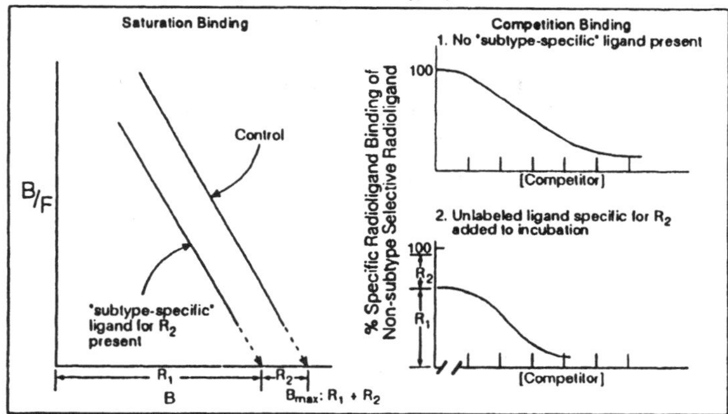

B. SUBTYPE-SELECTIVE (but not "specific") LIGANDS AVAILABLE

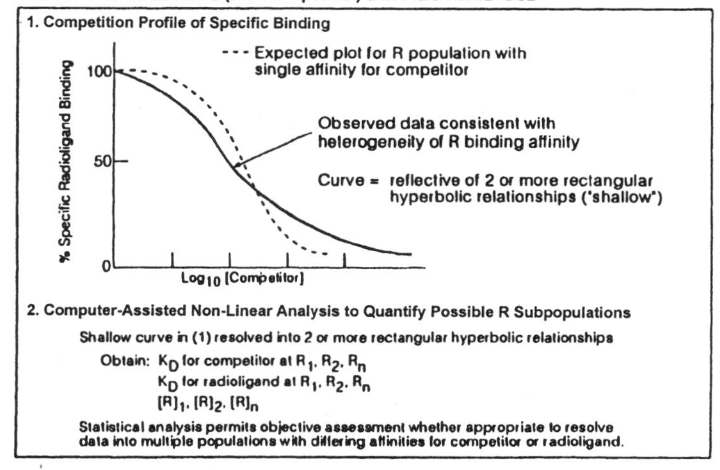

Figure 4-1. Two approaches to characterizing and quantitating receptor subtypes.
A. In the rare circumstances where an unlabeled but subtype-specific ligand is available, binding of a nonsubtype-selective radioligand can be evaluated in the absence and presence of the subtype-specific ligand, and information regarding the density of receptor subtypes obtained by "subtraction."

B. Typically, only subtype-selective competitors are available. In this situation, competition for the binding of a radioligand, which need not have identical affinity at the subtypes being evaluated, by a subtype-selective competitor is performed. The advantages of computer-assisted analysis of complex competition binding data are outlined in the text.

K_D values at R_1 and R_2 (assuming an *F* test supports a model with two sites) and relative $B_{\max}$ values (% of total receptor population as R_1 versus R_2) for each receptor subtype. An empirical test of this strategy for quantifying receptor subtypes that has helped determine the limits of computer-assisted analysis in this setting was undertaken by mixing known proportions of receptor subtypes in radioligand binding incubations. These studies revealed that accurate estimates of receptor subtype densities and affinities for the subtype-selective ligands could be obtained using a competitor that was only five- to eightfold selective for a particular receptor subtype when the subtypes were present in a 50:50 mix. The accuracy of binding parameter estimates for the two receptor populations in a 50:50 mix was, however, predictably improved as the competitor subtype selectivity increased to $\geq$ 50-fold. The practical limit of the ability to statistically resolve two receptor populations from one population was reached with a 90:10 mix of receptor subtypes. When such a small fraction of the total binding site population was contributed by one subtype, greater competitor selectivity was required, such that the competitor needed to possess an affinity constant 70- to 200-fold greater at one receptor subtype than at another for the statistical analysis inherent in the computer modeling program to favor a two-site fit over a one-site fit for the data (Hancock et al. [1979]). This same conceptual approach could be applied readily to mixtures of cells heterologously expressing one versus another subtype (or subtypes) of cloned receptors, and could be a useful empirical test of the feasibility of accurately characterizing mixtures of subtypes in native tissues for receptor sub-populations of interest.

Independent Data Consistent with the Existence of Receptor Subtypes

Even when rigorous analysis of complex binding phenomena has been performed and two or more receptor subtypes are described in a quantitative fashion, how can the investigator be confident that these putative independent receptor subpopulations actually exist and account for the complex binding phenomena observed in native biological preparations? First, studies of calculated receptor densities in cells or tissues with presumed mixtures of independent receptor subtypes should be independent of the subtype-selective competitor of radioligand binding used to reveal the existence of multiple receptor populations, and whether that competitor is an agonist or an antagonist. A variety of biological expectations might also be developed as data consistent with the existence of independent receptor subtypes being the explanation for complex binding phenomena in particular preparations. Fore example, independent receptor sub-populations would be expected to be expressed at differing ratios in different tissues and at different times during development. For receptors that have been molecularly defined, reverse transcription-PCR analysis of RNA from the tissue can document messenger

RNA expression of the receptor(s) of interest, and proteomic strategies document expression of the proteins coinciding with the presumed receptor subtypes contributing to the complex radioligand binding data detected.

The following discussion emphasizes differences in expectations for radioligand binding data that represent interconvertible receptor affinity states when compared with the above findings for discrete, independent receptor populations (or subtypes).

AFFINITY STATES OF A SINGLE RECEPTOR POPULATION

Interpretation of complex binding phenomena in terms of receptor subtypes implies that discrete macromolecules exist with differential selectivity for various ligands, and that these discrete receptor populations bind ligand independently of each other. However, a number of receptor systems have demonstrated complex radioligand binding phenomena that do not meet these criteria of discrete, non-interconvertible receptor populations. One example is the existence of receptor affinity states for agonist agents at GPCRs. Agonist-stabilized affinity states have been described for virtually all receptor populations linked to their effector systems via heterotrimeric GTP-binding proteins. The general observation is that agonist competition curves for radiolabeled antagonist binding to isolated membrane preparations are shallow. However, addition of GTP or guanine nucleotide analogs-agents essential to linking receptor occupancy to changes in effector activity-results in two fundamental changes in the agonist competition profiles. First, the curves shift to the right, i.e., the EC_{50} for agonist competition increases. Second, the shape of the agonist competition curves in the presence of guanine nucleotides is of normal steepness, in contrast to the shallow competition curves observed in the presence of agonist alone. Although for some receptor populations, these effects of guanine nucleotides are observed solely on receptor-agonist interactions, in many systems there are qualitatively (but not quantitatively) reciprocal effects of guanine nucleotides on receptor-antagonist interactions (e.g., Burgisser et al. [1982]). These reciprocal effects of guanine nucleotides on receptor-antagonist interactions appear to reflect a property of inverse agonists at these receptors, as will be discussed later.

Findings reported for β-adrenergic receptors linked to adenylyl cyclase stimulation will be used as an example of how the analysis of complex binding phenomena due to interconvertible affinity states has evolved with the acquisition of greater molecular understanding of these systems,. Agonist competition profiles for radiolabeled antagonist binding to β-adrenergic receptors are shallow and are modulated by guanine nucleotides. The ability of guanine nucleotides to increase incrementally both the EC_{50} for agonist competition and the steepness of competition profiles in a concentration-dependent manner indicates that the apparent heterogeneity of agonist binding is *not* a reflection of discrete, non-interconverting receptor populations (Kent et al. [1980]; DeLean et al. [1980]). That the degree to which guanine nucleotides modulate receptor-ligand interactions correlates with the intrinsic activity of the ligand in stimulating adenylyl cyclase activity is consistent with this conclusion. Competition for radiolabeled antagonist binding by so-called "full" agonists at β-adrenergic receptors is significantly influenced by guanine nucleotides; small shifts in EC_{50} values are noted for partial agonists, no influence of guanine nucleotides on receptor-antagonist interactions can be detected for "null" antagonists, and reverse effects of guanine nucleotides are observed for inverse agonists when competition for binding is examined in the absence (control) or presence of guanine nucleotides (Burgisser et al. [1982]; Samama et al. [1993]). The above observations indicate that the receptor subpopulation expressing high affinity receptor-agonist interactions is variable in nature, depending on the intrinsic activity of the ligand and the concentration of guanine nucleotide present.

The evolution of the molecular models contemplated to explain the existence of receptors with multiple affinity states are summarized in Table 4-1. Each model results in certain predictions for the binding properties that would be observed for receptor-agonist versus receptor-antagonist interactions. Model 1 describes the binding of drug D with two independent classes of noninteracting receptors, R and R', and is analogous to the model described above for analyzing receptor subtypes. The equilibrium association constants, shown as K and K', are assumed to be equal for classical competitive antagonists, since antagonists are not observed to discriminate between the two affinity states (or sites). However, K' is postulated to be higher than K for agonists, and agonists preferentially bind to the higher affinity form of receptor R' to yield HR'. This model would explain the distinct binding properties of agonists and antagonists observed at β-adrenergic receptors, but would not explain the different proportions of higher (R') and lower (R) affinity receptor forms that are noted in the presence of agonists with differing intrinsic activities or in the presence of varying concentrations of guanine nucleotides. (The proportion of receptors in the high-affinity state is denoted as % R_H in table 4-1 and throughout the subsequent text.) Although this mathematical model is insufficient to explain the observed interconvertibility of agonist affinity states, it is nonetheless suitable for obtaining parameters such as K_H (or K') and K_L (or K) for the high- and low-affinity state for agonists, respectively, under experimental conditions where the % R_H is unchanging (see DeLean et al. [1980]; Wregett and DeLean [1984]).

Table 4-1. Models of drug-receptor interactions that might account for ligand-induced interconvertible affinity states and expected binding properties for agents with differing efficacy.

Models		*Binding Properties*			
		Agonists		*Antagonists*	
		slope factor	**$\%R_H$**	**slope factor**	**$\%R_H$**
1. *Two noninterconvertible sites ($K' > K$ for agonists)* $D + R \overset{K}{\rightleftharpoons} DR$ $D + R' \overset{K'}{\rightleftharpoons} DR'$		<1	constant	1	none
2. *Cyclic (allosteric) model ($K' > K$ for agonists)* $D + R \overset{K}{\rightleftharpoons} DR$ $\updownarrows \quad \updownarrows$ $D + R' \overset{K'}{\rightleftharpoons} DR'$	2.1 at equilibrium:	1	none	1	none
	2.2 before equilibrium:	<1	small (<50%)	1	none
3. *Ternary complex model (TCM) ($L > 0$ for agonists)* $D + R \overset{K}{\rightleftharpoons} DR + X$ $M \updownarrows \quad \updownarrows L$ $D + RX \underset{K'}{\rightleftharpoons} DRX$		<1	varies with agonist	1	none
4. *Expansion of the TCM* $D + R + G \overset{K}{\rightleftharpoons} DR + G$ $J \updownarrows \quad \updownarrows \beta J$ $D + R^* + G \overset{\beta K}{\rightleftharpoons} DR^* + G$ $M \updownarrows \quad \updownarrows \alpha M$ $D + R^*G \overset{\alpha\beta K}{\rightleftharpoons} DR^*G$		<1	varies with agonist	≤1	varies with antagonist

5. Cubic Ternary Complex (CTC) Model

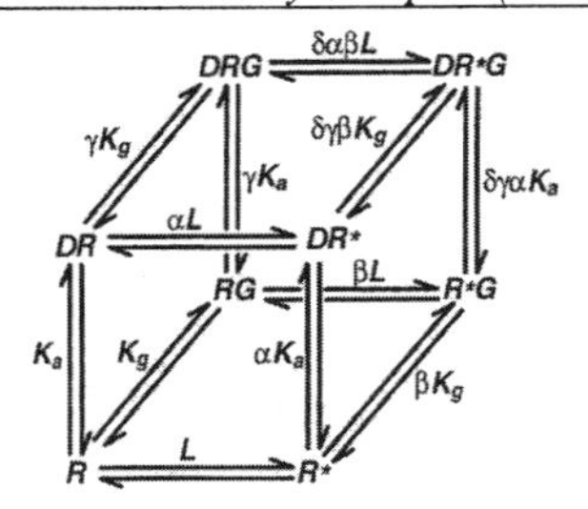

Note: K = equilibrium association constant, M^{-1}; % R_H = % of the receptor population manifesting a higher affinity for ligand. Table adapted from DeLean et al. (1980); models expanded to include $R \rightleftharpoons RX$ from Wreggett, K.A. and DeLean, A. (1984). Model 4 from Samama et al. (1993), where R* is a conformationally activated state of the receptor. The term K' in Model 3 is formally equivalent to αK in Model 4. Model 5, the cubic ternary complex model (Weiss et al. [1996a-c]) allows the inactive R to interact with G protein and active state. This model is formally identical to the allosteric two-state model of Hall (2000) developed to describe binding of an orthosteric versus allosteric agent to a receptor that exists in active or inactive conformations.

Model 2 is the "cyclic" model originally suggested by Katz and Thesleff (1957) for receptor-ligand interactions at the nicotinic cholinergic receptor. This model assumes that the receptor spontaneously exists in two freely interconvertible forms, denoted as R and R'. Antagonists are postulated to bind to both forms indiscriminately; the agonist preferentially binds to the R' form. The binding steps are assumed to be fast compared to the rates of isomerization between the two forms of the receptor, R and R'. The model predicts two categories of binding phenomena: those observed before equilibrium and those observed at equilibrium. Before equilibrium, agonist competition curves would be shallow, but only if the transition from DR to DR' is extremely slow relative to the binding steps, so that most of the ligand-receptor complex accumulates in the lower affinity form, DR. At equilibrium, the cyclic model is formally equivalent to an allosteric model for a monomeric receptor. The allosteric model would predict saturation and competition curves of normal steepness at equilibrium, with only one apparent form of the receptor interacting with agonist ligands. The predictions for Model 2 do not correlate with observations for agonist interactions at β-adrenergic receptors, since full agonists appear to interact with a high affinity "state" of the receptor that comprises a major fraction (≥ 50%) of the receptor population. Furthermore, the apparent heterogeneity of receptor-agonist interactions is independent of assay duration, counter to the predictions of Model 2 that apparent complexity of binding will disappear at equilibrium.

The Ternary Complex Model (TCM) and Expansions of the TCM

Model 3 introduces another membrane component, denoted *X*, into the molecular model for receptor-agonist interactions. In this model, transition from a low-affinity receptor state to a higher affinity state corresponds to a molecular transition from a *DR* complex to a ternary complex with the *X* component, denoted as *DRX*. Equations describing the ternary complex model are formally equivalent to those previously described for a "floating receptor model," where agonist occupancy of a receptor was postulated to elicit a conformation change in the receptor that results in more stable (or frequent) receptor encounters with effector units in the fluid mosaic of the membrane (Boenaems and Dumont [1977]; DeHaen [1976]; Jacobs and Cuatrecasas [1976]). The ternary complex model is a more general description of receptor-effector interaction. The unique ability of agonists to promote or stabilize the higher affinity receptor state would correspond to a large equilibrium association constant αM (dubbed *L* in early models, such as DeLean et al. [1980]) for transformation of *DR* to *DRX*. The full model shown in table 4-1 allows for the spontaneous occurrence of *RX* in the membrane in the absence of any ligand, *D* (Samama et al. [1993]). Spontaneous formation of the *RX* complex is determined by an equilibrium constant, *M*. Classical null antagonists bind to either receptor form with the same affinity ($K_{antag} = K'_{antag}$) and consequently do not stabilize the ternary complex ($M = \alpha M$ for antagonists). In contrast, for agonists K'_{ag} is greater than K_{ag}, and agonists stabilize the ternary complex ($\alpha M > M$ for agonists). It can be seen that the ratio of the equilibrium association constants (K'/K) equals the ratio $\alpha M/M$, which can be considered as not only a stability ratio for the *DRX* ternary complex, but also a measure of agonist efficacy in this system.

Several features of the ternary complex model made it an attractive candidate to explain the complexity of receptor-agonist interactions at G protein-coupled receptors. First, this model accounts for the observation that the proportion of receptors manifesting a high affinity state for agonist is variable, depending on the particular agonist studied, the G-protein content of the biological preparation, and the relative stoichiometry of *R* and *G* (equivalent to the "tissue component" of efficacy, introduced by Furchgott; cf. chapter 1). Thus, variability in the proportion of receptors in the *DRX* versus *DR* affinity state would be governed by the value of αM, which could vary. This model also accommodates the reciprocal effects of guanine nucleotides on receptor-agonist versus receptor-antagonist interactions observed in some target systems if $K' < K$ for antagonists, although in this case it must be noted that the "slope factor" for antagonists *also* would be < 1.0 (Burgisser et al. [1982]). Finally, this cyclical version of the ternary complex model could account for "precoupling" of receptors and G-proteins, occurring without prior exposure to agonist, noted by rigorous kinetic study of some systems (Neubig et al. [1988]).

Although Model 3 does not rely on any assumptions regarding the identity of membrane component *X*, the foregoing discussion implicates a GTP-binding regulatory protein as the probable component *X*, since interconversion between the two affinity states is controlled by guanine nucleotides. Several lines of independent biochemical evidence suggest that component *X* is a heterotrimeric G protein. First, agonist occupancy of G protein-coupled receptors stabilizes the formation of an agonist (Ag)-receptor-G protein complex that is resistant to detergent solubilization and can be characterized by its molecular radius as well as the identity of the G protein, using bacterial toxins or G protein-selective antibodies. Thus, the existence of physically definable Ag•R•G complexes is consistent with the ternary complex model (Limbird et al. [1980]; Michel et al. [1981]; Smith and Limbird [1981]; Kilpatrick and Caron [1983]).

The properties of *radiolabeled agonist* binding to GPCR also support the properties of receptor-agonist interactions evaluated in competition (denoted *antagonist/agonist) and direct (*agonist) binding studies analyzed based on the ternary complex model (data from DeLean et al. [1980]). Radiolabeled antagonist binding of ^{3}H-dihydroalprenolol (DHA) to β_2-adrenergic receptors in frog erythrocyte membranes suggested the existence of a single β-adrenergic receptor population. Competition for ^{3}H-DHA binding by the agonist hydroxybenzylisoproterenol (HBI) indicated that receptor-agonist interactions were heterogeneous, since modeling of these data yielded a slope factor for the competition curve < 1.0 (actually 0.83) and a fit of the data to a ternary complex model for the receptor, where 92% of the receptor population was in the high-affinity form and 8% was in the low-affinity form. The equilibrium association constant calculated for the high-affinity form (K_H) was 1.6×10^9 M^{-1} and for the low-affinity form (K_L) 2.2×10^7 M^{-1}. Since the K_L equilibrium association constant of 2.2×10^7 M^{-1} corresponds to an equilibrium dissociation constant for the lower affinity complex of 5×10^{-8} M, one can predict that agonist binding isolated by vacuum filtration will not be able to "trap" binding to this lower affinity form (see table 3-2). Consequently, one would anticipate that direct radiolabeled ^{3}H-HBI binding would selectively identify the higher affinity *DRX* ternary complex, and therefore predict that the density of ^{3}H-HBI binding sites would be approximately 92% of that detected by the antagonist ^{3}H-DHA, and that the affinity observed for ^{3}H-HBI binding should correspond to the K_H calculated from competition binding profiles. These predictions were indeed met with experimental data (DeLean et al. [1977]). Additional predictions can also be made. Since antagonists are not observed to discriminate between affinity states in this system, antagonist potency in competing for ^{3}H-DHA and ^{3}H-HBI binding should be virtually indistinguishable. In contrast, agonist competition for ^{3}H-HBI agonist binding should reflect interaction with the higher affinity receptor state. Consequently, competition profiles of *agonist/agonist curves would be predicted to be of normal steepness, since only a single affinity state is predicted to be identified by *agonist binding. In addition, the K_D calculated for an agonist from *agonist/agonist competition profiles should correspond to the K_D calculated for the high-affinity receptor state by computer resolution of the complex *DHA antagonist/agonist competition profiles. This prediction also was met by the data. A final prediction for radiolabeled agonist binding in this system is that *agonist binding should be modulated by guanine nucleotides. Addition of guanine nucleotides to steady-state incubations simultaneously with *HBI would be predicted to prevent detection of specific radioligand binding, and addition of guanine nucleotides to preformed *HBI-receptor complexes would be expected to accelerate their dissociation if guanine nucleotide-provoked changes in receptor affinity are due (at least in part) to changes in the rate constant for ligand dissociation, since ↓ affinity ≡ ↑ K_D ≡ ↑ k_{off}/k_{on} (cf. chapter 3). Again, this prediction was met by the data (DeLean et al. [1977]).

As studies of the β-adrenergic and other G-protein-coupled receptors proceeded, additional insights into their structure and function(s) were obtained that were not accounted for in the ternary complex model. One such property is the constitutive activation of some G protein-coupled receptors, first observed by Anolterz (1989) for δ opioid receptors in NG108-15 cells. Antagonists with negative intrinsic activity reduce "basal," or agonist-independent, activity in these systems. Overexpression of cloned receptors in heterologous cells increases the sensitivity of detection of constitutive receptor activity (Samama et al. [1993]) and, hence, an assessment of inverse agonist properties of antagonist drugs (Kenakin [2004a]).

Several properties of constitutively active GPCRs, particularly the G-protein-*in*dependent increased affinity for agonist, not accounted for by the ternary complex model (TCM), are accounted for by an extended version of the TCM that includes an isomerization of R to R^* (Table 4-1, Model 4). In this model, the structural component X in the ternary complex is explicitly defined as a heterotrimeric G-protein, G (Samama et al. [1993]). Here, the equilibrium association constants K and M are *unconditional*, and describe bimolecular reactions that are independent of each other. In contrast, αK and αM are *conditional* constants, as they describe the binding of a third component and are interdependent. The α term indicates how much the binding of D to R affects binding of G to R and vice versa. As indicated by Samama et al. (1993), α is a dimensionless coupling constant describing molecular efficacy in producing the active DR^*G ternary complex. The important addition of the extended TCM is the recognition that the receptor R can exist in two conformations, R and R^*. If it is assumed that only R^* can bind G-protein, then R^* represents the active state, and the equilibrium constant J describes the spontaneous isomerization of R to R^*. The β term, which is dimensionless, describes the extent to which the binding of D to receptor perturbs the $R \rightleftharpoons R^*$ equilibrium.

The extended TCM (Table 4-1, Model 4) is actually the most parsimonious modification of the ternary complex model that can account for all of the data for constitutively active receptors. It can be seen that if M (the equilibrium association constant describing $R^* + G \rightleftharpoons R^*G$) is a very small value or $G = 0$ (i.e., no G-protein is present or guanine nucleotides are in the incubation, preventing accumulation of R^*G), then the extended TCM becomes formally equivalent to the allosteric model for monomeric receptors proposed previously (Karlin [1967]; Thron [1973]; Colquhoun [1973], and discussed in chapter 1). Similarly, if J (the equilibrium constant for spontaneous isomerization of R to R^*) approaches a very high value, then *all* of the R are in the R^* state, and the extended TCM contracts to the initially proposed ternary complex model. The model also accounts for earlier observations that precoupling of adrenergic receptors to their cognate G-proteins occurs in certain target membranes (Neubig et al. [1988]), as agonist-independent formation of R^*G is actually predicted by the extended TCM, and its magnitude is governed by M. The ability of agonist to facilitate formation of the DR^*G complex (defined by α) also is a ratio of the affinities of D for R^* versus R^*G, accounting for earlier observations of the correlation between a drug's intrinsic activity and the ratio of K_H/K_L (DeLean et al. [1980]). Another important outcome of this model is that the affinity of D for R in the absence of G (described by K), is *not* related to the α coupling constant, consistent with classical observations that affinity and intrinsic activities of drugs are *not* correlated.

Constitutive activity, as noted above, is readily noted when GPCRs are over-expressed in heterologous systems. This empiric observation is consistent with mass action theory if receptors can spontaneously adopt an active state, R^*, independent of agonist occupancy (Kenakin [2001]):

$$R \overset{L}{\rightleftarrows} R^* + G \overset{\beta K}{\rightleftarrows} R^*G \tag{4.26}$$

where L is the allosteric constant defining the ratio of R to R^*, and βK is the affinity of the active receptor state, R^*, for the G protein, where $\beta > 1$ (cf. Table 4.1).

Equation 4.26 provides the possibility of agonist-independent constitutive activity, where the stoichiometry of the reactants, R^* and G, defines the extent of response in the absence of agonist . Thus, increasing receptor density, as occurs in over-expression systems, increases the probability of R^* so that:

$$\text{constitutive activity} = \frac{\beta L[R]K_{D_{R^* \cdot G}}}{T\beta L[R]K_{D_{R^* \cdot G}}} \quad (4.27)$$

where $K_{D_{R^* \cdot G}}$ is the equilibrium dissociation constant of the receptor (R^*)-G protein complex, βK in equation 4.26, and the values β and L are defined as in Table 4-1. Constitutive activity also is enhanced when G protein concentration is increased in a heterologous system, or by alteration of L, through means such as the introduction of point mutations to make receptors more constitutively active (cf. Kjelsberg [1992] and Samama et al. [1993]). A thermodynamically complete model for G protein-coupled receptor systems also has been developed, the Cubic Ternary Complex (CTC) Model (Table 4-1, Model 5). Naturally, the thermodynamically complete CTC model is inherently more complex, with more parameters than can be estimated on available tools for experimental observation. The critical addition to the CTC model is *ARG*, the non-signaling Ag•R•G protein complex. There are some experimental data consistent with the existence of non-signaling *ARG* complexes, including the inverse agonist properties of ICI 174,864 at the δ-opioid receptors (Chiu et al [1996]), and "cross-over" inhibition of insulin and insulin-like growth factor receptors by inverse agonist treatment of the cannabanoid CB_1 receptor (Boulaboula [1997]), presumably by G protein "trapping" in the *ARG* complex. Mutation of the α_{1B} adrenergic receptor to create a structure with increased affinity for agonist but reduced activation of the G_q-coupled phospholipase C (Chen et al. [2000]) has been interpreted to result from sequestering G_q in a conformation that promotes or stabilizes higher affinity binding of the agonist, but not a conformation (or conformation cycle) that supports coupling to phospholipase C. Such a mutation would be consistent with the existence of a non-signaling α_{1B} adrenergic receptor-G_q complex. A phenotypically similar mutation in rhodopsin underlies one allelic form of retinitis pigmentosa, emphasizing that some naturally occuring mutations manifest a non-signaling *ARG* state. Predictions of the *ETC* and *CTC* models differ quantitatively (Kenakin [2000]; Christopoulos and Kenakin [2002]) but not qualitatively. Given the parsimony of the *ETC* model, it is likely of greater use, except in systems where the non-signaling *ARG* complex is revealed to play a role.

The CTC, however, is of considerable value in describing allosteric regulation of G protein-coupled receptors where X (or G), does not represent an interacting GTP-binding protein, but rather where the second regulatory site is instead a binding site for an allosteric modulator of receptor binding and/or function (Christopoulos and Kenakin [2002]). Given the potential therapeutic utility of allosteric modifiers, especially allosteric enhancers, this use of the *CTC* model to predict properties of these allosteric systems will be invaluable in discerning the functional consequences of allosteric modulators of potential therapeutic significance. Finally, it should be appreciated that the TCM, ETC and CTC Models are so-called linkage models, i.e. they pre-define the species present in "thermodynamic space." Onoran et al. (1997; 2000), however, have developed models of GPCR behavior that exploit a probabilistic model. This model assumes that the receptor exists not in a particular state, but rather in a distribution of conformational states, and that binding of ligands, G proteins or allosteric regulators changes the distribution of receptor conformation states such that some are enriched and others are depleted. As emphasized by Kenakin [2004b], a probability model is more versatile than linkage theory models because it can predict receptor behavior beyond a single response, such as G protein activation, since GPCR are known to interact with a variety of proteins that impact receptor scaffolding, trafficking, and signal sensitivity (Brady and Limbird [2003]). Future studies will reveal the value of linkage versus probabilistic models of GPCR systems both in analyzing data and in predicting receptor behaviors that can be assessed in experiments that distinguish among discrete molecular hypotheses.

SUMMARY

This chapter summarized the different mathematical descriptions of complex binding phenomena that have been applied to radioligand binding data. These mathematical models are the basis for numerous computer programs now available for analyzing complex radioligand binding data. The quantitative parameters obtained from computer-assisted analysis, however, can be presumed to estimate K_D and B_{max} values for receptor-ligand interactions *only* when there is independent biochemical evidence that the mathematical model used for data analysis accurately reflects the molecular model describing the interaction of the ligand with its receptor(s). Two examples of molecular situations manifest by complex binding phenomena, i.e., the existence of independent receptor subtypes and interconvertible receptor affinity states due to receptor-G-protein coupling, were discussed in detail to provide the reader with a rational strategy for analyzing complex radioligand binding data and for obtaining the additional complementary data necessary to assure that the mathematical model used for analysis accurately reflects the biological model under study.

The emphasis on the ternary complex model and on expansion of that model was not intended to imply that all receptor-agonist interactions in all receptor systems necessarily will manifest interconvertible affinity states. Nor is there any reason to predict that multiple receptor affinity states (if they exist) will result from a ternary complex with transducer or effector molecules, as occurs in G-protein-coupled receptor systems. The discussion was meant to underscore how several independent experimental approaches can corroborate or refute any postulated explanation of heterogeneous (i.e., complex) receptor-ligand interactions. Models that predict observations different from those actually obtained in experimental studies can be excluded as explanations for the biological system under study, and should not be inherent in any mathematical algorithm used to obtain quantitative parameters for the binding data obtained. Similarly, models can be expanded to incorporate newly discovered properties of receptor systems, and thus provide algorithms for more rigorous data analysis. The important didactic aspect of this discussion, however, is not the findings or models per se, but the encouragement to combine a number of independent experimental approaches and analytical methods to document the internal consistency of the biochemical and radioligand binding data and the appropriateness of the postulated model that describes them.

REFERENCES

General

Boeynaems, J.M. and Dumont, J.E. (1977) The two-step model of ligand-receptor interaction. Mol. Cell Endocrinol. 7:33-47.

Christopoulos, A. and Kenakin, T. (2002) G Protein-Coupled Receptor Allosterism and Complexing. Pharm. Rev. 54:323-374.

DeLean, A., Hancock, A.A. and Lefkowitz, R.J. (1981) Validation and statistical analysis of a computer modeling method for quantitative analysis of radioligand binding data for mixtures of pharmacological receptor subtypes. Mol. Pharmacol. 21:5-16.

DeLean, A., Munson, P.J. and Rodbard, D. (1978) Simultaneous analysis of families of sigmoidal curves: Application to bioassay, radioligand assay and physiological dose-response curves. Am. J. Physiol. 235:E97-E102.

DeLean, A., Stadel, J.M. and Lefkowitz, R.J. (1980) A ternary complex model explains the agonist-specific binding properties of the adenylate cyclase-coupled β-adrenergic receptor. J. Biol. Chem. 255:7108-7117.

Janin, J. (1973) The study of allosteric proteins. Prog. Biophys. Mol. Biol. 27:77-119.

Kenakin, T.P. (2001) Quantitation in Receptor Pharmacology. Rec. & Chan. 7:371-385.

Kenakin, T. (2004a) Efficacy as a Vector: The Relative Prevalence and Paucity of Inverse Agonism. Mol. Pharmacol. 65:2-11.

Kenakin, T. (2004b) Principles: Receptor Theory in Pharmacology. Trends Pharm. Sci. 25:186-193.

Klotz, I.M. (1946) The application of the law of mass action to binding by proteins: Interactions with calcium. J. Am. Chem. Soc. 9:109-117.

Klotz, I.M. and Hunston, D.L. (1975) Protein interactions with small molecules: Relationships between stoichiometric binding constants, site binding constants, and empirical binding parameters. J. Biol. Chem. 250:3001-3009.

Koshland, D.E., Nemethy, G. and Filmer, D. (1966) Comparison of experimental binding data and theoretical models in proteins containing subunits. Biochem. 5:365-385.

Limbird, L.E. and Motulsky, H. (1998) Receptor Identification and Characterization. In *Handbook of Physiology: The Endocrine System* (vol. 1). *Cellular Endocrinology*, chapter 4. Oxford: Oxford University Press, Inc.

Motulsky, H. and Christopoulos, A. (2003) GraphPad Prism. Fitting models to biological data using linear and non-linear regression. A practical guide to curve fitting. San Diego: GraphPad Software Inc. (see www.graphpad.com).

Munson, P.J. (1983) LIGAND: A computerized analysis of ligand binding data. Methods in Enzymol. 92:543-546.

Munson, P.J. and Rodbard, D. (1980) LIGAND: A versatile computerized approach for characterization of ligand-binding systems. Anal. Biochem. 107:220-239.

Newsholme, E.A. and Start, C. (1973) *Regulation in Metabolism*, chapter 2. New York: John Wiley and Sons.

Onaran, H.O. et al. (2000) A look at receptor efficacy. From the signaling network of the cell to the intramolecular motion of the receptor. In *The Pharmacology of Functional, Biochemical, and Recombinant Systems Handbook of Experimental Pharmacology* (vol. 148), pp. 217-280. New York: Springer-Verlag.

Onaran, H.O. and Costa, T. (1997) Agonist efficacy and allosteric models of receptor action. Ann. NY Acad. Sci. 812:98-115.

Rodbard, D. (1973) A graphic method for the determination and presentation of binding parameters in complex systems. Anal. Biochem. 20:525-532.

Samama, P., Cotecchia, S., Costa, T. and Lefkowitz, R.J. (1993) A mutation-induced activated state of the β_2-adrenergic receptor: Extending the ternary complex model. J. Biol. Chem. 268:4625-4636.

Steinhardt, J. and Reynolds, J.A. (1969) *Multiple Equilibria in Proteins*, chapter 2, pp. 10-33. New York: Academic Press.

Teipel, J. and Koshland, D.E. (1969) The significance of intermediary plateau regions in enzyme saturation curves. Biochem. 8:4656-4663.

Weiss, J.M., Morgan, P.H., Lutz, M.W. and Kenakin, T.P. (1996a) The cubic ternary complex receptor-occupancy model. I. Model description. J. Theor. Biol. 178:151-167.

Weiss, J.M., Morgan, P.H., Lutz, M.W. and Kenakin, T.P. (1996b) The cubic ternary complex receptor-occupancy model. II. Understanding apparent affinity. J. Theor. Biol. 178:169-182.

Weiss, J.M., Morgan, P.H., Lutz, M.W. and Kenakin, T.P. (1996c) The cubic ternary complex receptor-occupancy model. III. Resurrecting efficacy. J. Theor. Biol. 181:381-397.

Wregett, K.A. and DeLean, A. (1984) The ternary complex model: Its properties and application to ligand interactions with the D_2-dopamine receptor of the anterior pituitary gland. Mol. Pharmacol. 26:214-227.

Specific

Adair, G.S. (1925) The hemoglobin system. VI. The oxygen dissociation curve of hemoglobin. J. Biol. Chem. 63:529-545.

Birdsall, N.J.M., Hulme, E.C. and Burgen, A.S.V. (1980). The character of the muscarinic receptors in different regions of the rat brain. Proc. Roy. Soc. Lond. B. 207:1-12.

Bouaboula, M., Perrachon, S., Milligan, L., Canatt, X., Rinaldi-Carmona, M., Portier, M., Barth, F., Calandra, B., Pecceu, F., Lupker, J., Maffrand, J.P., LeFur, G. and Casellas, P. (1997) A selective inverse agonist for central canabinoid receptor inhibits mitogen-activated protein kinase activation stimulated by insulin or insulin-like growth factor 1. Evidence for a new model of receptor/ligand interactions. J. Biol. Chem. 272:22330-22339.

Brady, A.E. and Limbird, L.E. (2002) G protein-coupled receptor interacting proteins: Emerging roles in localization and signal transduction. Cellular Signal. 14:297-309.

Burgisser, E., DeLean, A. and Lefkowitz, R.J. (1982) Reciprocal modulation of agonist and antagonist binding to muscarinic cholinergic receptors by guanine nucleotide. Proc. Natl. Acad. Sci. USA 79:1732-1736.

Bylund, D.B., Blaxall, H.S., Iversen, L.J., Caron, M.G., Lefkowitz, R.J. and Lomasney, J.W. (1992) Pharmacological characteristics of alpha$_2$-adrenergic receptors: comparison of pharmacologically defined subtypes with subtypes identified by molecular cloning. Mol. Pharmacol. 42:1-5.

Colquhoun, D. (1973) The relation between classical and cooperative models for drug action. In *Drug Receptors*, H.P. Rang (ed.). Baltimore: University Park, pp. 149-182.

DeHaen, C. (1976) The non-stoichiometric floating receptor model for hormone sensitive adenylyl cyclase. J. Theoret. Biol. 58:383-400.

Feldman, H.A. (1972) Mathematical theory of complex ligand-binding systems at equilibrium: Some methods for parameter fitting. Anal. Biochem. 48:317-338.

Hancock, A.A., DeLean, A.L. and Lefkowitz, R.J. (1979) Quantitative resolution of beta-adrenergic receptor subtypes by selective ligand binding: application of a computerized model fitting technique. Mol. Pharmacol. 16:1-9.

Jacobs, S. and Cuatrecasas, P. (1976) The mobile receptor hypothesis and "cooperativity" of hormone binding: Application to insulin. Biochim. Biophys. Acta 433:482-495.

Karlin, A. (1967) On the application of a "plausible model" of allosteric proteins to the receptor of acetylcholine. J. Theoret. Biol. 16:306-320.

Katz, B. and Thesleff, S. (1957) A study of the "desensitization" produced by acetylcholine at the motor end plate. J. Physiol. 138:63-80.

Kent, R.S., DeLean, A. and Lefkowitz, R.J. (1980) A quantitative analysis of beta-adrenergic receptor interactions: Resolution of high and low affinity states of the receptor by computer modeling of ligand binding data. Mol. Pharmacol. 17:14-23.

Kilpatrick, B.V. and Caron, M.G. (1983) Agonist binding promotes a guanine nucleotide reversible increase in the apparent size of the bovine anterior pituitary dopamine receptors. J. Biol. Chem. 258:13528-13534.

Kjelsberg, M.A., Cotecchia, S., Ostrowski, J., Caron, M.G. and Lefkowitz, R.J. (1992) Constitutive activation of the alpha 1B-adrenergic receptor by all amino acid substitutions at a single site: Evidence for a region which constrains receptor activation. J. Biol. Chem. 267:1430-1433.

Klotz, I.M. (1983) Ligand-receptor interactions: What we can and cannot learn from binding measurements. Trends in Pharm. Sci. 4:253-255.

Klotz, I.M. and Hunston, D.L. (1984) Mathematical models for ligand-receptor binding: Real sites, ghost sites. J. Biol. Chem. 259:10060-10062.

Lavin, T.N., Hoffman, B.B. and Lefkowitz, R.J. (1981) Determination of subtype selectivity of alpha-adrenergic ligands. Comparison of selective and non-selective radioligands. Mol. Pharmacol. 20:28-34.

Limbird, L.E., Gill, D.M. and Lefkowitz, R.J. (1980) Agonist-promoted coupling of the β-adrenergic receptor with the guanine nucleotide regulatory protein of the adenylate cyclase system. Proc. Natl. Acad. Sci. USA 77:775-779.

Michel, T.M., Hoffman, B.B., Lefkowitz, R.J. and Caron, M.G. (1981) Different sedimentation properties of agonist- and antagonist-labeled platelet $alpha_2$-adrenergic receptors. Biochem. Biophys. Res. Comm. 100:1131-1134.

Neubig, R.R., Gantzos, R.D. and Thomsen, W.J. (1988) Mechanism of agonist and antagonist binding to α_2-adrenergic receptors: Evidence for a precoupled receptor-guanine nucleotide protein complex. Biochemistry 27:2374-2384.

Smith, S.K. and Limbird, L.E. (1981) Solubilization of human platelet α-adrenergic receptors: Evidence that agonist occupancy of the receptor stabilizes receptor-effector interactions. Proc. Natl. Acad. Sci. USA 78:4026-4030.

Thron, C.D. and Waud, D.R. (1968) The rate of action of atropine. J. Pharm. Exp. Ther. 160:91-105.

5. THE PREPARATION AND STUDY OF DETERGENT-SOLUBILIZED RECEPTORS

Some properties of cell surface receptors cannot be determined in sufficient detail without purification of the putative receptor and subsequent reconstitution of its biological function(s). As a first step in receptor isolation, the investigator must remove the receptor from the biological membrane so that the receptor can be isolated based on its own physicochemical properties, rather than on those of the membrane as a whole. The typical experimental approach for solubilizing the receptor from the membrane is to use biological detergents, agents whose physical properties resemble those of the lipid constituents of the membrane bilayer. The receptor is thus lured from the membrane into detergent micelles and can then be studied as a unique biochemical entity.

In this chapter, the physical properties of biological membranes will be reviewed briefly as an introduction to the properties of biological detergents that make them suitable for solubilization of a membrane-bound receptor. The choice of a biological detergent for accomplishing certain goals will then be discussed. Finally, methods appropriate for assessing binding to a detergent-solubilized preparation will be summarized.

GENERAL PROPERTIES OF BIOLOGICAL MEMBRANES AND DETERGENT MICELLES

Cell surface receptors are embedded in a plasma membrane which is composed of proteins as well as phospholipids, glycerolipids, and cholesterol. Phospholipids compose the major fraction of the membrane lipid mixture, and the physical properties of the membrane are, for the most part, a reflection of the properties of the constituent phospholipids.

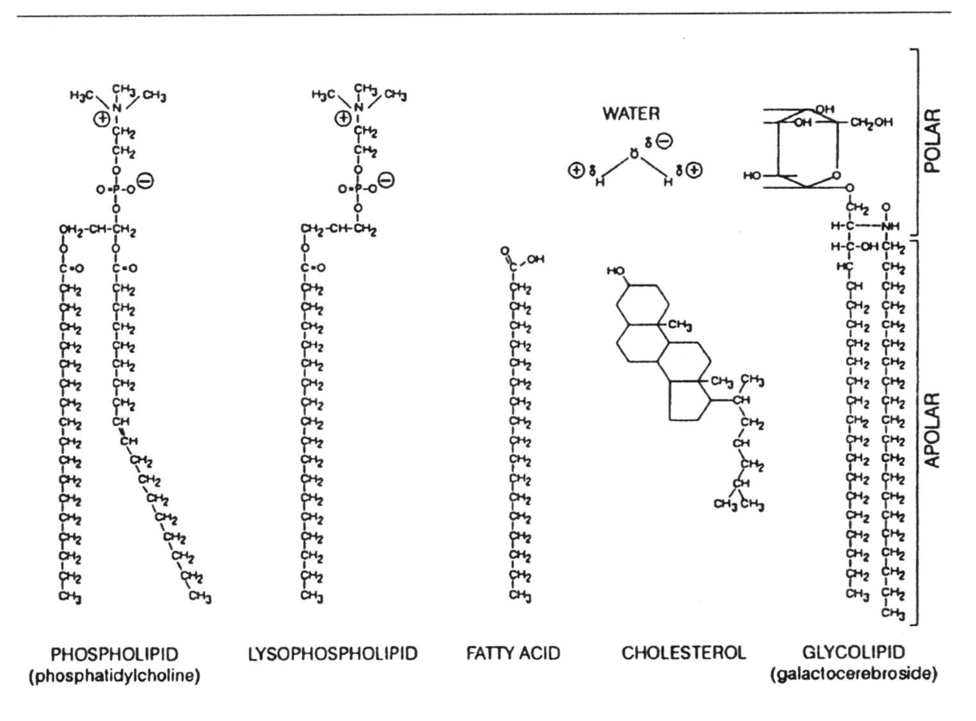

Figure 5-1. The structures of the amphipathic components of biological membranes.

As shown in figure 5-1, phospholipids (like glycerolipids and, perhaps less strikingly, cholesterol) are **amphipathic** in structure. This means that a portion of the phospholipid molecule is hydrophilic ("water loving," polar in nature) and a portion of the molecule is hydrophobic ("water hating," apolar in nature). Phospholipids are organized into bilayers in the aqueous environment characteristic of biological systems. These bilayers are arranged such that the nonpolar "tails" of the lipid are sequestered from the aqueous environment to form a hydrophobic core, while the hydrophilic "heads" project to form a polar surface on each side of the bilayer (figure 5-2).

Organization of phospholipids into a bilayer, rather than a micelle, results from the need to bury two hydrophobic tails per amphipathic molecule, and is thus dictated by packing considerations. The bilayer structure permits phospholipids to attain the lowest free energy state because the polar head groups are available to interact with polar water molecules (cf. figure 5-1); the apolar fatty acid moieties are effectively protected from the polar, aqueous environment. These bilayers spontaneously seal to form vesicular structures for the same thermodynamic reasons; if "edges" existed, the apolar core of the bilayer would be forced to deal (in a thermodynamic way) with the polar water (see Tanford [1973;1978]).

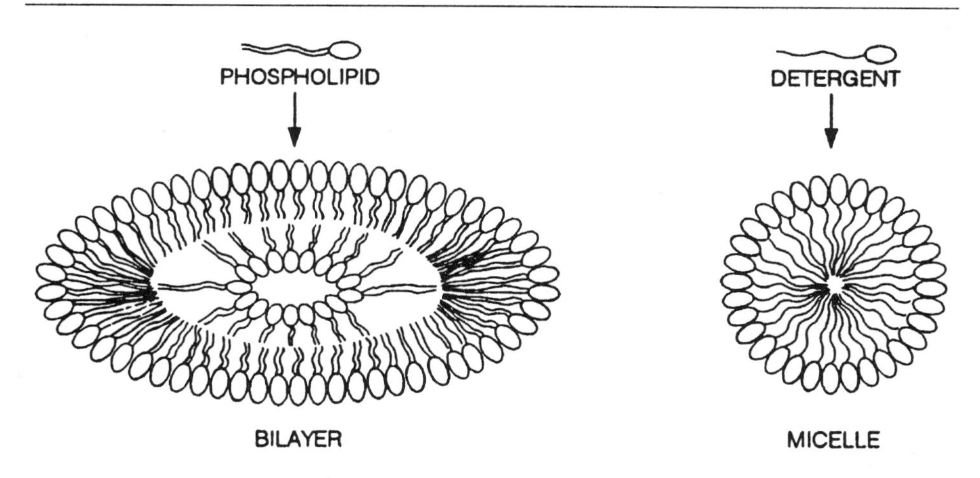

Figure 5-2. Assembly of amphipathic molecules into bilayers or micelles.

Figure 5-3 provides a schematic diagram of the arrangement of proteins in the biological membrane. Membrane proteins can be divided operationally into either **peripheral** (extrinsic) or **integral** (intrinsic) proteins. The association of integral proteins with the membrane can be defined further by virtue of their penetrance through to the extracytoplasmic face of the membrane (**ectoproteins**) or not (**endoproteins**) (cf. figure 5-3). Transmembrane-spanning ectoproteins can be assumed to have carbohydrate moieties covalently attached to one or more amino acid side chains. Peripheral proteins are associated with the membrane via electrostatic interactions at the polar face of the bilayer and thus often can be dissociated from the membrane by mild treatments, such as high ionic strength or chelating agents. Integral proteins extend across the lipid bilayer as one or more α helices or, in some cases, extended β sheets, interacting with both the polar and apolar regions of the lipid bilayer. Consequently, hydrophobic bond-breaking agents, such as biological detergents or chaotropic agents, are required to dissociate integral proteins from the bilayer. Following extraction into biological detergents, integral proteins often are still associated with membrane lipids and become insoluble or aggregated when lipids or the solubilizing detergent are removed (for broad overviews concerning membrane proteins, see Steck [1974]; Bretscher [1985]).

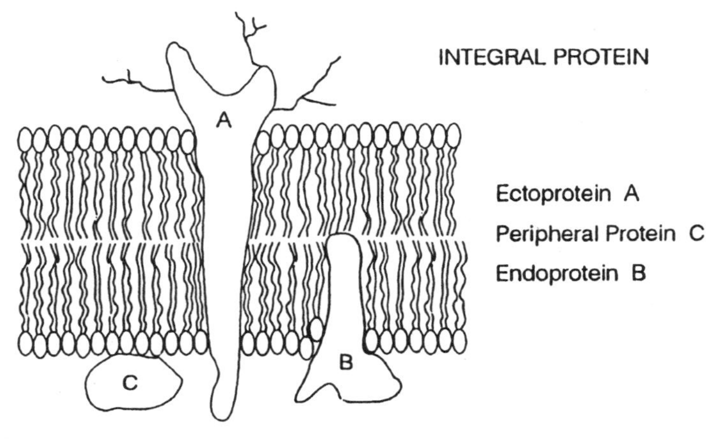

- Ectoproteins may span the membrane multiple times (polytopic) or once (monotopic).
- Proteins also may be attached to the exofacial surface via lipid linkages, such as glycophosphatidyl inositol (GPI) anchors.

Figure 5-3. Proteins can associate with the membrane bilayer in several possible ways.

Because detergent micelles mimic the amphipathic nature of the membrane, biological detergents can effectively disrupt the biological membrane and persuade integral proteins to leave the membrane bilayer and associate with a detergent micelle. Biological detergents contain a polar sphere attached to a single apolar hydrophobic tail and thus resemble fatty acids or lysophospholipids. As mentioned above, detergents tend to associate into micelles, rather than bilayers, due to thermodynamic packing considerations. In micelles, the hydrophobic tails of detergents are effectively buried from water, and the polar head groups are accessible at the micelle surface for interaction with water. A schematic structure of a micelle, and its distinction from a bilayer, is shown in figure 5-2.

Formation of a micelle from detergent monomers is a cooperative effect and occurs over a very narrow concentration of monomer, usually referred to as the **critical micelle concentration** (**cmc**). Figure 5-4 demonstrates that as the concentration of detergent added to an aqueous solution increases, a concentration of monomer is reached beyond which the addition of increasing detergent concentrations results in micelle formation. This detergent concentration is the critical micelle concentration. An important relationship is demonstrated in figure 5-4: at concentrations of detergent greater than the cmc, the concentration of monomer in solution is equal to the cmc. The cmc of a detergent often turns out to be the most critical factor to consider when selecting a detergent for a particular experimental use, because the cmc dictates both the concentration of detergents that must be added to solubilize integral proteins (as solubilization is believed to be accomplished by micellar forms of the detergent) and the effectiveness of different methods for subsequent detergent removal. For example, since detergent micelles cannot freely pass through dialysis membranes with pore sizes small enough to exclude smaller molecular weight proteins, the rate of detergent removal by dialysis depends on the concentration of detergent monomer. Consequently, dialysis will be an effective procedure for removal or exchange of detergents that possess relatively high cmc values, but will be of little use for detergents with low cmc values, because an insufficient driving force for diffusion across the dialysis membrane (dictated by the monomeric detergent concentration) will exist.

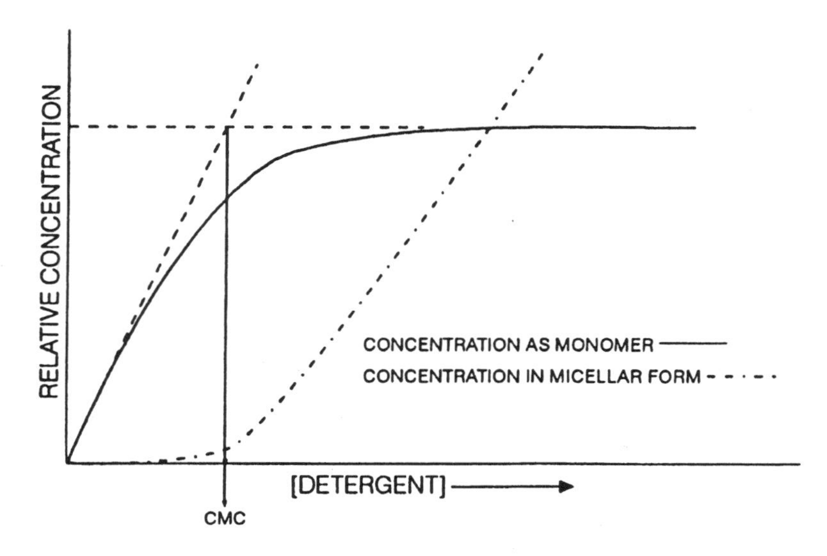

Figure 5-4. The relationship between the detergent concentration added to an aqueous solution and the concentration of monomer or micellar species in solution.

Certain principles can be applied to predict whether a detergent will have a relatively high or low cmc. In detergents with apolar chains of similar length, ionic detergents will have approximately a 100-fold higher cmc than nonionic detergents. This is a consequence of two phenomena. First, the association of ionic amphipathic compounds into micelles is counteracted by electrostatic repulsion from other detergent monomers possessing a similar net charge. Second, ionized headgroups "bind" water, creating an effectively larger headgroup and counteracting the thermodynamic forces that favor micelle formation. The cmc of ionic detergents decreases as the ionic strength of the aqueous medium increases, because the increased concentration of counter ions in the aqueous medium decreases electronic repulsion of the charged headgroups. In contrast, the cmc of nonionic detergents is affected very little by the ionic strength of the aqueous medium. Since the cmc of a detergent varies with differing buffer conditions, it is often necessary to determine the cmc for a detergent under the experimental conditions where it will be used. (For examples of methods for determining cmc values, see Colichman [1951]; Jain and Wagner [1980]; Rosenthal and Koussale [1983].)

The cmc of detergents is affected to some extent by the aggregation number of detergents, i.e., the mean number of detergent monomers per micelle. The aggregation number is influenced by both the apolar and polar moieties of the detergent molecule.

One feature that may determine whether or not an integral protein will be soluble in a particular detergent is the micelle diameter, because the diameter dictates the span of hydrophobic surface that is offered to accommodate the hydrophobic surface of an integral protein. The diameter of a micelle cannot be greater than two times the length of the alkyl chains that comprise the monomer. Since the alkyl chains are normally in a liquid state under the conditions of micelle formation, the effective alkyl chain length is actually less than that for fully extended chains, ranging from approximately 24Å for alkyl chains that are 12 carbons in length to approximately 30Å for alkyl chains 16-18 carbons in length (see Tanford [1973]). Thus, the diameter of the hydrophobic core of a micelle composed of a detergent with a C_{16}-C_{18} alkyl chain is similar to that of the hydrophobic core of the biological membrane, which is also approximately 30Å. Additional thermodynamic calculations demonstrate that detergent micelles are probably not spherical, as shown in figure 5-2B, but instead are disk-like with rounded ends, and possess an optimal surface area/head group of approximately 60-65Å, which corresponds to an aggregation number of 100 monomers/micelle composed of C_{12} monomers (Tanford [1973]; Helenius and Simons [1975]).

Table 5-1 summarizes the structures and properties of commercially available biological detergents that are most widely used for solubilization and characterization of integral membrane proteins, including cell surface receptors. The detergents are organized based on their structural similarity with phospholipids, cholesterol, or glycerolipids. Since fatty acids and lysophospholipids form micelles when introduced into aqueous solutions, these agents also can be considered biological detergents. Several characteristics of the detergents are variable when comparing the detergent structures summarized in table 5-1, including:

1. Size of the polar head group;
2. Presence (ionic) or absence (nonionic) of a *net* surface charge on the polar headgroup;
3. Length of the apolar chain;
4. Presence of ring structures as the source of hydrophobicity;
5. Presence or absence of sugar moieties.

Differences in structural features of detergents account for differences in aggregation number and, most importantly, in the cmc values for different detergents. General characteristics of each group of detergents are summarized below. Short chain phospholipids also can serve as biological detergents (Hauser [2000).

Lysophospholipids and fatty acids are not commonly used for the solubilization of cell surface receptors from native membranes. This may be due to their high cost (if used in pure form) to prepare in solutions at and above their cmc. Detergents that resemble lysophospholipids in their overall structure are available in ionic and nonionic forms. Ionic detergents often are not used for solubilization of receptors in a functional state for two fundamental reasons. First, ionic detergents are often significantly more deleterious to biological activity than nonionic detergents with comparable hydrophobic moieties; for example, sodium dodecyl sulfate (structure, table 5-1) is known to denature proteins. Second, ionic detergents are not suitable in situations where ion-exchange chromatography is anticipated, as in solubilization of a receptor with the intent to purify the molecule, because interaction with ion exchange resins will be dictated primarily by the net charge of the detergent, which will obscure subtle differences in the net charges of detergent-associated proteins that one is attempting to fractionate. The nonionic detergents of the Lubrol, Brij, Triton, and Tween series have been demonstrated to be very effective in solubilizing membrane proteins in high yields. Furthermore, it has been observed empirically that addition of sucrose (e.g., 0.25 M) or glycerol (5-30%) increases both the yield of membrane protein extraction and the retention of protein function subsequent to solubilization.

Table 5-1. **Representative biological detergents.**

Category/Structure	Chemical Name	Trade Name	CMC	Aggregation No.
Lysophospholipids				
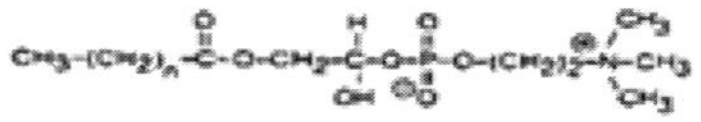	Lysolethicin		10 μM (C_{16}) - 8 mM (C_{10})	181 (C_{16})
Structural Analogs of Lysophospholipids				
Anionic 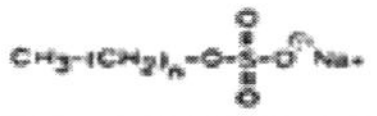	Sodium Dodecyl (or Lauryl) Sulfate	SDS	2 - 8 mM (in H2O)	62
Cationic 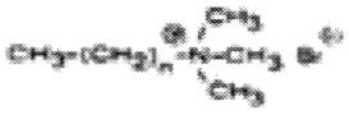	Cetyl Trimethyl Ammonium Bromide	CTAB		169
Nonionic 	Polyoxyethylene Alcohol	Brij. Series, Lubrol W. Al. Series	2 μM	
	Polyoxyethylene-O-para-Octyl Phenol	Triton-X Series Nonidet P40	0.01-0.02% for TX-100 (avg. MW = 288)	140
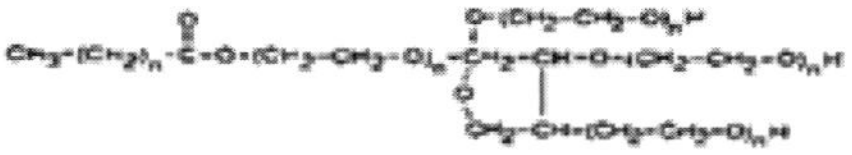	Polyoxyethylene Sorbital	Tween Series	0.0013% (Tween 80)	60
Structural Analogs of Cholesterol				
Anionic 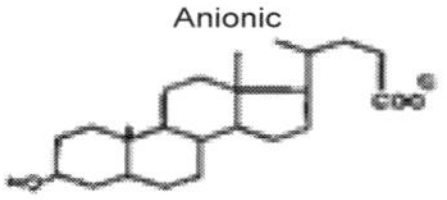	Cholate		10 mM (in H2O)	2-4
Nonionic 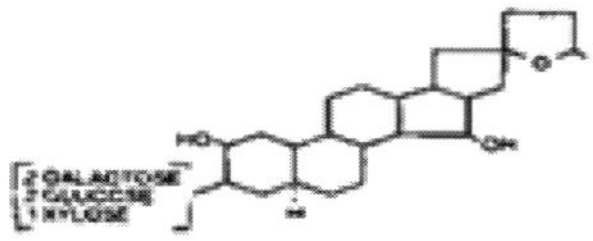	Digitonin		0.75 mM or 0.08% w/vol (MW 1200)	60
Zwitterionic 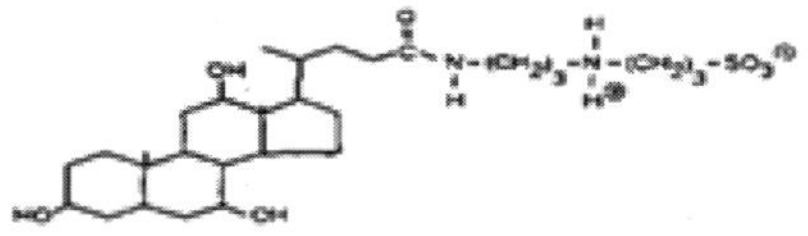	1-(3-Cholamide-Propyl)-Dimethyl Ammonio-1-Propane Sulfonate (N-Alkyl-Sulfobetaines)	CHAPS	4-6 mM	
Structural Analogs of Gylcolipids				
Alkyl Glucosides 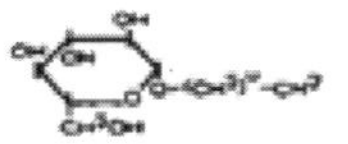	Octyl-β-D-Glucoside (when n=7)		25 mM (n=7; cmc ↓ with ↑ alkyl chain length)	
β-D-Maltoside 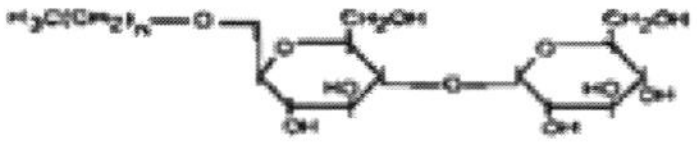	Dodecyl-β-D-Maltoside	(when n=11)	100-600 μM (not accurately known)	98
N-D-Gluco-N-Methylalkanamide 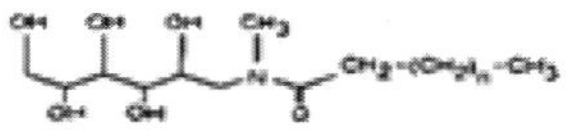	Omega (Octanoic Acid, n=5); Mega-9 (Nonanoic Acid, n=6); Mega-10 (Decanoic Acid, n=7)			

The values for cmc and aggregation number were obtained from A. Helenius et al. (1979), *Methods in Enzymology* 56:734-749 and references therein.

It is wise to obtain the Brij, Lubrol, Triton and Tween series from a manufacturer that provides them at ultra-high purity, because heavy metal and oxidant contaminants of "reagent grade" preparations often can inactivate the biological functions of proteins extracted into the detergent-solubilized preparation (see Ashani and Catravas [1980]; Chang and Bock [1980]). Even when these nonionic detergents are chemically pure, commercial products of the polyoxyethylene series (i.e., O-CH_2CH_2-containing polar groups) are almost always heterogeneous due to a broad distribution of polyoxyethylene chain lengths. Consequently, the value of *n* shown in table 5-1 (when provided by the manufacturers) is actually a number average. The hydrocarbon part of these detergents is also heterogeneous, but to a lesser extent than the polar polyoxyethylene groups. Triton series detergents absorb in the ultraviolet region due to the presence of the phenyl ring. Tween 80, a detergent of the polyoxyethylene sorbitol series, also has absorption bands at 268 and 279 nm, which give Tween 80 its yellow color (Helenius et al. [1979]). The absorbance by Triton and Tween in the 280 nm range interferes with protein determinations that rely on measuring absorbance in the ultraviolet region.

Several biological detergents that possess a cholesterol-like backbone are available. In contrast to the detergents discussed above, bile-salt detergents that resemble membrane cholesterol in their chemical structure possess no clear-cut polar domain. Instead, the polar groups are distributed along the length of the molecule, thus making one side of the molecule polar. These molecules do not associate into micelles in the manner shown schematically in figure 5-2. Instead, the bile salts are proposed to associate into detergent particles by association of the polar faces of 2-8 molecules in a manner that resembles the overlapping of cardboard box lids (see Helenius and Simons [1975], figure 4). Yields of solubilization with the anionic detergent cholate, usually prepared as a sodium salt of cholic acid, typically are improved by solubilization in the presence of high concentrations of salt, e.g., 0.5 M NaCl or KCl. Purification of commercial cholate preparations, if not provided in ultra-pure form, often is advisable to protect biological activities in the detergent-solubilized preparation from inactivation by cholate contaminants (Ross and Schatz [1976]). Digitonin is a cholesterol-resembling nonionic detergent prepared from plant extracts, with trace impurities that vary from batch to batch. Such impurities have been observed to alter significantly the yield of solubilization of intrinsic proteins and their biological functions subsequent to solubilization. Contaminants present in digitonin also alter the solubility of the detergent. Solutions of digitonin are prepared by dissolving the digitonin powder in just-boiled water and then 10-100X buffer stock solutions to achieve the desired final buffer concentrations; with time, digitonin-containing solutions will form some precipitation. Digitonin-containing buffers used in sucrose gradients or in column-eluting solutions should be prepared several days in advance, the contaminants allowed to precipitate, and the buffer filtered through 0.45-micron filters before use. Although digitonin operationally can be purified by letting an aqueous solution sediment for several days at 4°C and then lyophilizing the nonprecipitated detergent, the resulting more purified material often is less effective in removing proteins from the membrane than the crude digitonin supplied by a manufacturer. Despite the frustrations and relatively high cost of digitonin, however, investigators have nonetheless observed that digitonin often solubilizes cell surface receptors with minimal perturbation of binding properties, particularly receptors of the seven transmembrane-spanning, G-protein-coupled structural family. This may be due to the fact that digitonin-protein micellar particles contain considerable membrane lipid when compared with detergent-protein complexes that result, for example, from Triton solubilization. It is possible that digitonin does not perturb the receptor microenvironment crucial for receptor binding activity.

Brown and Schonbrunn (1993) have determined that the somatostatin receptor (a member of the G-protein-coupled, seven transmembrane-spanning receptor structural family) can be quantitatively extracted (85% yield) in association with G-proteins using dodecyl β-maltoside (Rosevear et al. [1980]) in a 4:1 (gram/gram) ratio with cholesteryl hemisuccinate. The precise nature of the micellar structure of this detergent:cholesterol:protein mixture is not known at present. Because these superior extraction properties are achieved with other receptors (Wilson and Limbird [2000]), this detergent mixture has superceded digitonin as a "gentle" detergent for preferred use in extraction of functional receptor molecules, particularly because of its lesser cost and greater chemical purity without batch-to-batch variability.

The zwitterionic detergent CHAPS, which also has a cholesterol-like backbone, seems to combine the best of all worlds. It is gentle in that it typically solubilizes receptors with their binding properties intact. Furthermore, CHAPS and its 2-hydroxy analog CHAPSO have a defined structure and are prepared by chemical synthesis, not by extraction from biological material. Unlike digitonin, these detergents do not vary among batches and do not precipitate from solution. CHAPS and CHAPSO also have a relatively high cmc, making them amenable to reconstitution studies using dialysis for lipid exchange and detergent removal. Finally, the zwitterionic headgroup of CHAPS (or CHAPSO) has no net charge over the pH range of 2-10, and thus CHAPS and CHAPSO are amenable to ion exchange chromatography.

Several commercial detergents resemble glycolipids in structure in the sense that the polar group contains a sugar moiety. Strictly speaking, Tween and other detergents possessing sorbitol moieties should also be considered in this group. Alkyl glucosides and the "Mega" series (e.g., Mega-9, table 5-1) share several operationally attractive features with CHAPS. They are structurally homogenous when supplied by the manufacturer; they are particularly gentle in terms of retention of biological activity in detergent solution, and they possess relatively high cmc values. Like CHAPS, the alkyl glucosides and Mega-9 detergents are enjoying increased use for solubilization and functional characterization of cell surface receptors (cf. Stubbes et al. [1976] for an example of rhodopsin extraction using octylglucoside).

CHOICE OF A BIOLOGICAL DETERGENT

Choosing a biological detergent for solubilization of a functional receptor often is entirely empirical in nature, i.e., one selects a detergent that successfully removes the receptor from the biological membrane in such a way that ligand-receptor interactions still occur in the membrane-free environment. (Methods for assessing ligand-receptor interactions in detergent solutions are summarized later.) When an investigator is fortunate enough to have a selection of several detergents that appear to successfully solubilize a functional receptor, a few detergent properties that might influence future experiments should be considered in identifying the "detergent of choice." Some of these features were indicated above in summarizing properties of the detergents shown in table 5-1.

Overall, nonionic detergents are more effective than ionic detergents in solubilizing an integral membrane protein with its biological activity intact. If receptor purification is the ultimate goal, selecting the detergent that gives the best yields of receptor solubilization would seem particularly wise, even if long-term receptor stability requires almost immediate exchange of the receptor into another detergent or a lipid mixture subsequent to solubilization. In some cases, even the *detection* of detergent-solubilized receptors requires dilution of the detergent to reduce the final detergent concentration (Smith and Limbird [1981]) or exchange of detergent-solubilized receptors into other detergents (Repaske et al. [1987]) or into lipid vesicles (Fleming and Ross [1980]). When the investigator anticipates that ion-exchange chromatography (or other separation procedures that resolve molecules based on net surface charge) will be utilized for receptor purification, the use of ionic detergents for receptor solubilization should be avoided, since receptors solubilized into micelles of ionic detergents will fractionate primarily based on the net charge of the detergent micelle rather than on chemical properties of the micelle-associated proteins. Thus, exchange of ionic for nonionic detergents would need to occur before performing non-exchange chromatography. If the ultimate goal of experimental studies is to characterize the hydrodynamic properties of the receptor molecule, two features of the detergent should be considered. First, the feasibility of using a detergent of the polyoxyethylene series (e.g., Lubrol) should be assessed, since the partial specific volume of Lubrol (like that of phospholipids) is in the range of 0.9-1.00 cm^3/g; detergent binding to the protein does not confuse interpretation of the physical parameters obtained using equilibrium sedimentation (Edelstein and Schachman [1967]; Reynolds and Tanford [1976]). In contrast, a detergent like digitonin or cholate has a partial specific volume close to that for proteins (partial specific volume 0.75-0.8 cm^3/g), and small differences in receptor conformation that result in even subtle differences in detergent binding will result in significant differences in M_r values calculated from sedimentation studies (Helenius et al. [1979]). Secondly, if the receptor will be monitored during these hydrodynamic evaluations by following absorbance of the protein, i.e., OD 280 nm, then neither Triton-X 100 nor Tween 80 should be selected as the solubilizing detergent, since these detergents are not optically clear at 280 nm. If experimental goals include ultimate reconstitution of receptor function or reassociation of proteins existing in separate detergent particles, detergent removal via dialysis may be a useful procedure to achieve these ends. As mentioned earlier, selection of a detergent with a relatively high cmc would be particularly wise, since the cmc is equal to the concentration of monomer in a micellar solution, which determines the driving force for dialysis.

An investigator frequently is faced with the reality that no one detergent may be optimally suited to accomplish each goal of a research project. Consequently, it often is necessary to exchange the detergent-solubilized protein from one detergent to another. When feasible, dialysis can be used, but this is a reasonable choice only for detergents possessing a high cmc value, where the monomer concentration (which can cross the dialysis membrane) is relatively high and thus creates sufficient driving force to effect detergent removal in a reasonable time frame. More rapid exchange can be accomplished by gel-filtration chromatography on desalting columns. For example, a protein-detergent complex in Triton X-100 could be exchanged into octylglucoside by gel filtration of the Triton X-100-solubilized protein on a column equilibrated with, and eluted in, an excess of octylglucoside. Removal of detergent tightly bound to the protein might require incubation at higher temperatures (18-37°C) and repeated gel-filtration steps. A similar approach can be used to exchange detergents during a purification scheme. These resins may be specific affinity chromatography resins developed by conjugating a receptor-specific ligand to a resin backbone or may be more general resins, such as streptavidin-agarose to bind receptor occupied by biotinylated ligand or an antibody-conjugated resin to bind native or epitope-tagged receptors. The detergent-solubilized preparation can be adsorbed to a particular resin in one detergent and eluted from that resin in another detergent. Regardless of the method chosen to accomplish detergent exchange, it is always essential to maintain the concentrations of all detergents above their respective cmc values in order to maximize exchange into other detergents, and especially to avoid aggregation or precipitation of the detergent-solubilized protein.

SOLUBILIZING RECEPTORS FROM BIOLOGICAL MEMBRANES

The above section discussed the rationale for selecting biological detergents based on the experimental goals the investigator has for the ultimate solubilized preparation. In reality, however, the most suitable biological detergent is one that is determined empirically to "work," and only one detergent may fulfill this minimum criterion. The effectiveness of a detergent is assessed not only by its ability to release the receptor molecule from the membrane but also by its ability to minimally perturb the receptor structure so that the receptor retains its characteristic specificity for particular ligands in the detergent-solubilized preparation.

A useful strategy for comparing several detergents of interest to the investigator is the following:

1. Select several detergents to compare, perhaps one or two detergents from each structural category. Prepare detergent solutions at concentrations equal to or greater than the cmc for the detergent, since it is believed that detergent micelles (rather than monomers) actually solubilize integral proteins. Besides, only micelles provide an amphipathic environment similar to the native bilayer environment that the integral protein prefers, and the environment is essential for retaining the receptor in solution.
2. Determine whether incubation with the detergent solution for various times (Lichtenberg et al. [1983]) and at various temperatures (e.g., 4°, 15° or 25°C) can release prelabeled ligand-receptor complexes into the supernatant following a 100,000 × g 60-minute centrifugation. Release into a 100,000 × g 60-minute supernatant is the minimum criterion for whether or not a receptor is truly solubilized (see later). If most of the prelabeled ligand-receptor complexes remain in the pellet, then the detergent or the incubation conditions are not sufficiently disruptive to release the receptor. If most of the radioactivity is in the supernatant of a 100,000 × g 60-minute centrifugation, but is free in solution rather than still bound to receptor, then detergent concentration or incubation conditions may be unsuitable for effective solubilization of the receptor in a ligand-bound or functional state. If most of the radioactivity is in the supernatant and is receptor-associated, then the detergent apparently succeeds in both releasing the receptor and in not interfering with

receptor-ligand interactions, or at least in not destabilizing those interactions that occurred prior to exposure to the detergent.

For those detergents that show promise in releasing receptor-ligand complexes from the membrane, the above pilot experiments are then repeated using membrane preparations that are not occupied by ligand at the time of solubilization. In this way, the solubilization protocol can be refined to optimize the detection of unoccupied receptor subsequent to solubilization using radioligand binding techniques.

When performing such exploratory experiments for evaluating detergent and incubation conditions for optimal receptor solubilization, it is important to document whether the receptor is in the pellet or the supernatant of a 100,000 × g centrifugation 60 minutes, i.e., to account for the recovery of all relevant biological material. If the receptor is not in either fraction, then the detergent or incubation conditions being evaluated may be inactivating the receptor. If the receptor does not appear to be solubilized but instead can be retrieved quantitatively in the pellet of a 100,000 × g 60-minute centrifugation, then the detergent being evaluated may simply need to be used at higher concentrations, or added to the membrane preparation in such a way that it might be more effective, e.g., with homogenization or at higher temperatures. When an investigator has found a detergent and incubation conditions that appear to solubilize the receptor effectively, the genuine solubility of the receptor preparation can further be documented by determining whether the solubilized preparation meets the following operational criteria for a membrane-free preparation:

1. Receptor should remain in the supernatant of a 100,000 × g 60-minute centrifugation;
2. Detergent-solubilized receptor should pass through a 0.22-micron filter;
3. Receptor preparations should not contain any membrane vesicles when evaluated using electron microscopy, and
4. Solubilized receptor should elute in the "included volume" of a gel permeation resin such as Sepharose 6B (molecular size for exclusion approximately 0.5-1.0 × 10^6 M_r). This latter criterion may be ignored if the protein under study is known to possess a molecular weight in excess of this range or a rigid molecular structure that would cause it to carve out a large sphere or "volume" of solution.

Occasionally, not all of these criteria can be met, even for a truly solubilized receptor preparation. For example, many detergent-solubilized receptors do not quantitatively pass through a 0.22-micron filter-not because the receptor is associated with particulate material, but because the receptor interacts with the filter through electrostatic or other chemical attractions. In this situation, however, the investigator can be more confident of the monodisperse nature of this same detergent-solubilized preparation if the receptor were demonstrated to elute in the included volume of gel-sieving columns and to be free of membrane vesicular structures when evaluated by electron microscopy. Finally, a monodisperse preparation should permit the ligand-binding moiety (presumably the receptor) to fractionate independently of the bulk of the protein in the detergent-solubilized preparation during subsequent gel-sieving, ion-exchange, hydrophobic and/or adsorption chromatography procedures.

METHODS FOR ANALYSIS OF DETERGENT-SOLUBILIZED RECEPTORS

Many methods used for assaying membrane-bound receptors, such as vacuum filtration and centrifugation, are not appropriate for detergent-solubilized receptors, at least without some modifications. As indicated above, truly solubilized receptors should pass through filters and should not sediment during microcentrifugation, and a number of approaches have been developed (some of them quite ingenious) for detection of binding to detergent-solubilized receptors. These will be summarized below.

Equilibrium Dialysis

Equilibrium dialysis is applicable to detergent-solubilized receptors if care is taken to select a dialysis membrane that is truly permeable to radioligand but impermeable to the detergent-solubilized receptor. Nonetheless, the same practical limitations cited in chapter 3 may preclude using equilibrium dialysis to measure binding to detergent-solubilized receptors. Its principal limitation is that it may be difficult to detect the cpm due to the ligand-receptor complex, since the cpm on one side of the dialysis membrane indicates free radioligand and cpm on the other side equals bound (B) + free (F) ligand. As a result, the investigator is usually measuring a small signal (B) over a large "background" (i.e., F).

Hummel-Dreyer Chromatographic Procedure

Another method for assessing binding to soluble or detergent-solubilized receptors which does not perturb equilibrium conditions is the chromatographic procedure developed by Hummel and Dreyer (1962), shown schematically in figure 5-5. This method quantitates the binding of a small ligand (e.g., *D*) to a protein or other macromolecule (*R*). A gel exclusion resin is selected that excludes *R* and includes *D*. A column is prepared with dimensions that permit complete resolution of *D* and *R*, and is then equilibrated with ligand *D*. The receptor-containing preparation is applied to the column in the same ligand-containing solution used to equilibrate and elute the column. As *R* moves through the column, *R* removes *D* from the solution within the gel. Subsequently, *R* and *DR* complexes elute from the column in the void volume to give a peak in the elution profile of *D*. This peak is followed later by a trough that extends to the included volume of the column. The trough is a consequence of the removal of *D* from the column by *R*. When equilibrium conditions prevail, the amount of *D* removed from solution by the receptor protein (the trough) is theoretically equal to the excess of *D* in the earlier peak (see Cann and Hinman [1976]).

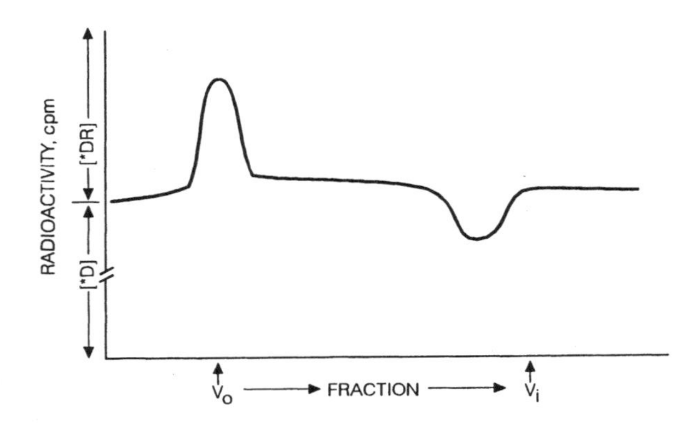

Figure 5-5. The Hummel-Dreyer chromatographic procedure for detecting the quantity of ligand binding to a detergent-solubilized receptor.

An advantage of the Hummel-Dreyer procedure is that there are two opportunities to measure the quantity of ligand bound to its receptors, i.e., either in the peak or the trough. This is particularly fortunate when *D* is monitored spectrophotometrically, because the receptor protein present in the peak might interfere with the absorption measurements. In this situation, binding is assessed by quantitating the area of the trough. Nonetheless, there are several potential disadvantages to the Hummel-Dreyer method when ligand binding to detergent-solubilized proteins would rely on radioactivity (i.e., **D*) to monitor the quantity of **DR* complexes. First, the method is potentially expensive, because the column and all equilibrating and eluting buffers must contain the radioligand **D*. Second, the concentration of **D* will be quite large compared to the concentration of *R*, and thus the baseline of cpm for **D* could easily obscure a small peak or trough in the column elution profile that represents the quantity of **D* bound to *R*. Generation of a detectable peak often requires that the receptor preparation be applied in high concentration, as was done for measurement of ^{3}H-GTP binding to tubulin by a modification of the Hummel-Dreyer approach (Levi et al. [1974]).

The limitation of a potentially small signal (**DR*) above a large "background" (**D*) for both the equilibrium dialysis and the Hummel-Dreyer procedure has caused investigators to be less concerned with whether equilibrium is perturbed when the concentration of bound ligand (i.e., quantity of **DR* complexes) is assessed and more concerned with developing methods that may permit the sensitive detection of detergent-solubilized receptors. Several methods for this purpose are discussed below.

Gel Filtration

Gel filtration of ligand-receptor complexes on resins suitable for desalting purposes is widely used for quantitation of ligand binding to detergent-solubilized receptor preparations. The procedure is summarized schematically in figure 5-6.

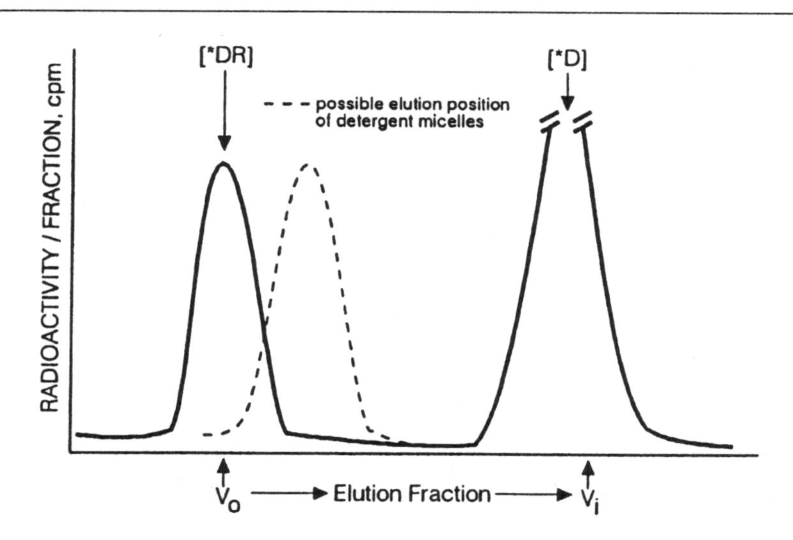

Figure 5-6. Gel filtration as a means for resolving receptor-ligand complexes from free ligand. The possible association of ligand with detergent micelles or mixed lipid-detergent micelles can result in high values for nonspecific ligand binding, but can be minimized if larger height-to-width ratios are selected for the column bed, thus permitting at least some resolution of the ligand-receptor-micelle particle from protein-free micelles (see text). V_o = void volume; V_i = included (or salt) volume.

The resin for the gel-filtration column is selected so that the receptor is excluded from the matrix and elutes in the "void volume" (V_o) of the column, whereas the ligand is included in the resin. Commonly, G-50 fine Sephadex resins are used for this purpose. Resolving bound from free ligand is relatively straightforward when the ligand is a small molecular weight drug or neurotransmitter. For larger ligands, such as polypeptide hormones, a resin may not be available that permits rapid and quantitative resolution of receptor-ligand complexes from free ligand. Gel-filtration methods can also be less suitable for resolving free ligand from ligand bound to detergent-solubilized receptors when there is substantial interaction of the ligand with detergent micelles. This is because micelles often elute, at least partially, in the void volume of a desalting column (see schematic illustration of this potential difficulty in figure 5-6). To determine whether this problem needs to be considered, the elution profile of a sample containing only ligand and detergent can be assessed to verify that receptor-ligand complexes can be resolved from ligand-detergent complexes. It also is wise to document the elution profile of receptor-ligand complexes from the column, rather than to assume that the elution profile of blue Dextran or other void-volume markers is a valid prognosticator of where the receptor-ligand complex of interest will elute. Occasionally, the hydrophobic nature of detergent-solubilized receptors can result in receptor adsorption to the resin matrix (which is also typically hydrophobic) and cause retention of the receptor-ligand complexes so that they elute later than the void volume defined by elution of blue Dextran. Increasing the concentration of detergent in the elution buffer often can decrease this hydrophobic adsorption. Regardless of the elution properties of the receptor and detergent micelles, however, so-called "desalting" columns for terminating binding incubations to detergent-solubilized receptors typically are "taller" (larger height-to-width ratio) than recommended by the resin manufacturer for desalting homogeneous aqueous solutions. This is to achieve at least some separation of the receptor-micelle preparation from protein-free micelles (see figure 5-6). A convenient column is a 5-ml disposable plastic pipette, plugged with siliconized glass wool and filled to a 4-ml capacity with Sephadex G-50 fine. Access to the column bed is facilitated by removing the narrower portion at the top using a heated single-edge razor blade to cut off the top of the pipette. Alternatively, spin columns of smaller dimensions can be utilized, and optimized, based on empirical success in resolving bound from free ligand and minimizing non-receptor background ligand binding to detergent micelles.

When gel filtration is used to terminate a binding incubation containing detergent-solubilized receptors, it is assumed that the association of ligand with receptor terminates as soon as the sample is applied to the column. Subsequent detection of ligand-receptor complexes is obviously dependent on *minimizing dissociation* of these complexes during the gel-filtration step and *optimizing recovery* of these complexes from the resin. Dissociation of ligand-receptor complexes is minimized by decreasing both the time and the temperature of the gel-filtration step. Rapid elution of ligand-receptor complexes is permitted when columns are as small as possible (for complete resolution of receptor from ligand) and are composed of a resin material that has very rapid flow rates. Occasionally, these small gel-filtration columns are fitted into or on top of centrifuge tubes and the entire assembly centrifuged so that the void volume can be quickly isolated into the centrifuge tube. Performing the gel filtration in a cold room (i.e., 6-10°C) decreases the possibility of ligand-receptor dissociation during the gel-filtration procedure. When the binding incubation is performed at higher temperatures than the ultimate gel-filtration step, it is advisable to cool the incubation before applying it to the resin. The recommendation of reduced temperature for gel filtration of ligand-receptor complexes, however, is only sensible if the receptor does not demonstrate a markedly reduced affinity for ligand at lower temperatures, in which case cooling the incubation might result in more rapid rather than less rapid dissociation of ligand from receptor.

Precipitation

Several methods have been developed that precipitate detergent-solubilized ligand-receptor complexes in such a way that ligand bound to detergent-solubilized receptor can be resolved from free ligand using centrifugation or vacuum filtration techniques analogous to those used for membrane-bound receptors. These methods include **polyethylene glycol** and **ammonium sulfate precipitation.** In both procedures, water molecules in the detergent-solubilized preparations become "organized" around polyethylene glycol (Atha and Ingham [1981]) or the ammonium sulfate salt (Green and Hughes [1955]), thus effectively reducing the solubility of the protein in the detergent-solubilized preparation. Incubations of radioligand with detergent-solubilized receptors can be terminated by adding polyethylene glycol or ammonium sulfate to a final concentration that has been determined empirically to precipitate solubilized receptor-ligand complexes. For polyethylene glycol, this is usually $\geq$ 8% final concentration. For ammonium sulfate, the concentration effecting precipitation is more variable, but 50% saturation is commonly employed. When ammonium sulfate is used, care must be taken to offset the decrease in pH that occurs as increasing concentrations of ammonium sulfate are added to the incubation medium. Polyethylene glycol or ammonium sulfate are added to the incubation after steady-state binding has been attained. An interval of time (minutes to hours) is then required to effect precipitation of the ligand-receptor complexes. Carrier proteins can be added at the same time as polyethylene glycol or ammonium sulfate to accelerate the precipitation process as well as improve the yield of receptor obtained in the precipitate. The precipitated material is separated from the solution by either filtration or centrifugation. Yields of precipitated detergent-solubilized receptors adsorbed to filters often are improved by presoaking the filters in polyethylene glycol or ammonium sulfate-containing solutions. Nonspecific binding is reduced by adding bovine serum albumin to these presoaking solutions and by washing the filters after filtration of the receptor preparation with additional volumes of polyethylene glycol or ammonium sulfate-containing solutions. When centrifugation is used to terminate the incubation, the free radioligand trapped in the pellet during centrifugation can be removed by repeated washings of the precipitate with an incubation buffer containing the same final concentration of polyethylene glycol or ammonium sulfate needed to precipitate the receptor.

A critical requirement for success of the precipitation approach in resolving receptor-bound ligand from free radioligand is that the free ligand must remain in solution under conditions required to precipitate the receptor. This is more of a potential problem when the ligands are polypeptides than when ligands are small molecular weight drugs or neurotransmitters. The ability to detect binding to the detergent-solubilized receptors using the precipitation approach also requires that the ligand-receptor complexes do not dissociate to a significant extent during the interval necessary to effect precipitation following polyethylene glycol or ammonium sulfate addition and isolating and washing receptor-ligand complexes following precipitation. (For an example of polyethylene glycol precipitation for assaying binding to detergent-solubilized receptors, see Brown and Schonnbrunn [1993]; for examples of ammonium sulfate precipitation, see Gorissen et al. [1981].)

Adsorption to Filters

Another approach to separating radioligand bound to detergent-solubilized receptors from free radioligand is to change the properties of filters typically used for vacuum filtration so that they bind and thus retain the detergent-solubilized protein. In this way, vacuum filtration can be used to resolve free ligand from ligand bound to detergent-solubilized receptors in a manner analogous to the methods used for study of membrane-bound receptors. An example of this approach is to treat glass fiber filters with polyethyleneimine (PEI) (Bruns et al. [1983]) or DEAE (Keith et al. [1982]) so that the filters possess a net positive charge. Because most detergent-solubilized receptor preparations possess a net negative charge at physiological pH, filtration of detergent-solubilized receptors through PEI-coated filters results in retention of the receptor material on the filter. When the investigator is fortunate, the free radioligand passes through the filter. Nonspecific binding can be reduced by washing the filter with incubation buffer. A similar conceptual approach using commercially available DEAE-cellulose filters is suitable when rapidly dissociating ligands are used to identify detergent-solubilized receptors.

Another approach for estimating the amount of radioligand-receptor complex that forms in a detergent-solubilized binding assay is to terminate the incubation by adding activated charcoal to assay tubes. Often it is found that radioligands, particularly those of small molecular weight, are rapidly adsorbed to the charcoal (i.e., within 0-5 minutes). After centrifugation in a microfuge to sediment the charcoal and adsorbed ligand, the receptor-bound ligand can be estimated by determining the radioactivity in the supernatant. Adsorption of receptor material to the charcoal can be minimized by prior coating of the charcoal with bovine serum albumin (Gorissen et al. [1981]) or dextran (Leff and Creese [1982]). Similarly, counts in the supernatant due to free ligand can be reduced by prewashing and gravity-sedimenting the charcoal to remove so-called "fines."

Receptor Immobilization

Another conceptual approach for detecting radioligand binding to detergent-solubilized receptors is to immobilize the receptor to a solid support. This conceptual approach also is applicable when the receptor can be immobilized in a variety of ways: lectin-conjugated agarose resins; biotinylation of receptors on the surface of intact cells prior to solubilization and subsequent adsorption to avidin-agarose; immunoisolation of receptors using anti-receptor antibody adsorbed to protein A- (or G-) agarose; or immunoisolation of epitope-tagged receptors. Two alternate experimental protocols are shown schematically in figure 5-7. In the first protocol (figure 5-7A), detergent-solubilized receptors are adsorbed to a lectin affinity resin based on the glycoprotein nature of the cell surface receptor prior to incubation with radioligand. Alternatively, as in figure 5-7A, receptor ligand complexes can be formed prior to resin-dependent resolution of bound from free radioligand. The incubation is terminated by diluting the incubation volume and washing the resin, using repeated centrifugation to remove the free radioligand. Radioligand bound is quantified in a scintillation (^{3}H) or gamma (^{125}I) counter, as appropriate. This experimental approach in figure 5-7A has been successfully applied to the study of solubilized receptors for epidermal growth factor (Nexo et al. [1979]. The approach in figure 5-7B has been used successfully for the hepatic glucagon receptor (Herzberg et al. [1984]) and transcobalamin II using concanavalin A-sepharose as the resin or α_2 adrenergic receptors using agarsose as the resin (Wilson and Limbird [2000]) as examples. Methods using ligand mobilization for interacting with solubilized receptor also have been described (Oka et al. [2004]), and are

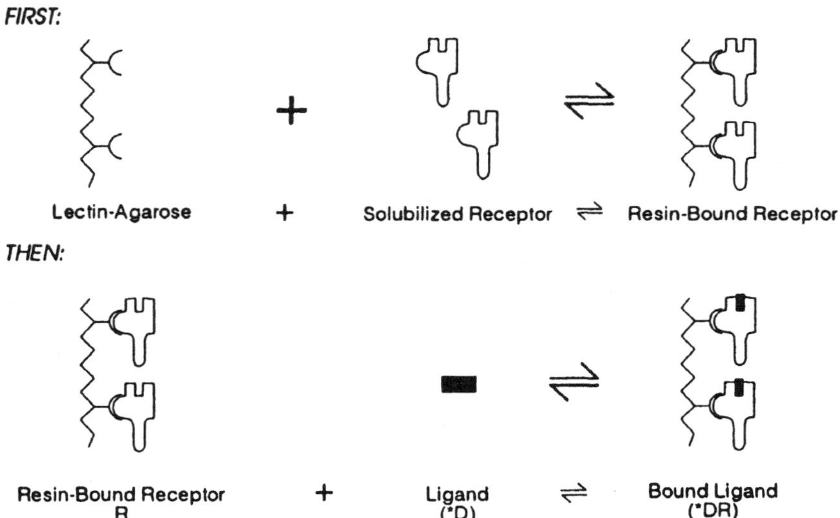

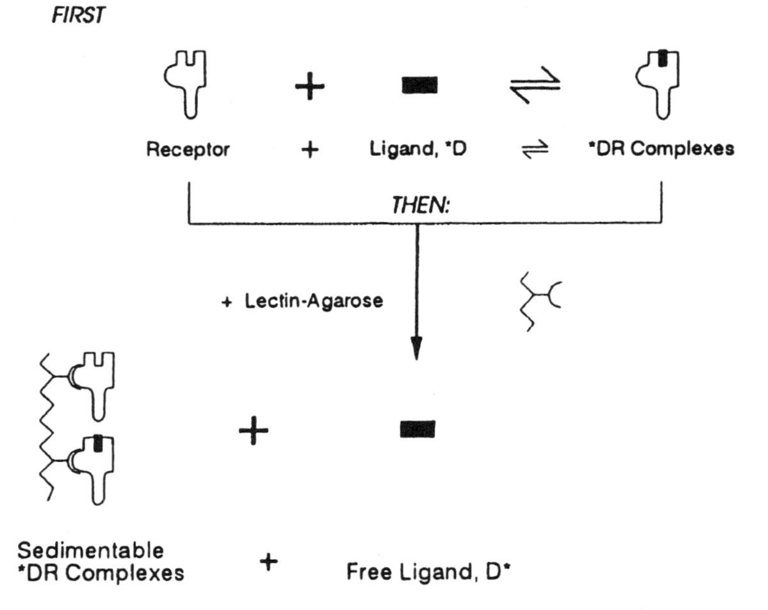

Figure 5-7. Immobilization of detergent-solubilized receptors to lectin-conjugated agarose as a tool for assaying binding to receptors in solution. The same principle can be applied to detecting binding to receptors immunoisolated via antibody and protein A- (or G-) agarose, or to receptors, biotinylated while on the cell surface of intact cells, adsorbed to avidin-agarose gel matrices.

appropriate for receptor identification or assessment of protein-protein interactions. However, quantitative analysis of binding using these procedures to yield K_D values, will still require that all of the assumptions inherent in the equations for analysis are met, e.g. $[R]$ is significantly less than K_D for ligand, such that ligand is in excess.

SUMMARY

This chapter discussed, from a pragmatic standpoint, some available approaches for preparing and assaying detergent-solubilized preparations of membrane-bound receptors. The guidelines for selection of a detergent and assay for detecting ligand-receptor complexes in solution are primarily empirical in nature. Except in extraordinary circumstances, nearly all methods for assaying binding to detergent-solubilized receptors will disturb the equilibrium of $*D + R \rightleftharpoons *DR$.. However, only the most fortunate investigator has the luxury of worrying about whether equilibrium is being perturbed when the binding assay is terminated. Most investigators simply are seeking a sensitive and reliable assay for detergent-solubilized receptor binding that

permits an assessment of whether the solubilized ligand-binding site possesses the specificity characteristic of the membrane-bound receptor. If the detergent-solubilized binding site meets the criteria expected for the physiological receptor of interest, then the optimized assay for detergent-solubilized binding will be useful for identification and characterization of the receptor during subsequent purification and reconstitution protocols.

REFERENCES

General

Bretscher, M.S. (1985) The molecules of the cell membrane. Sci. Amer. 253:100-109.

Hauser, H. (2000) Short-chain phospholipids as detergents. Biochim. Biophys. Acta 508:164-181.

Helenius, A., McCaslin, D.R., Fries, E. and Tanford, C. (1979). Properties of detergents. Methods in Enzymol. 56:734-749.

Helenius, A. and Simons, K. (1975) Solubilization of membranes by detergents. Biochim. Biophys. Acta 415:24-79.

Lichtenberg, D., Robson, R.J. and Dennis, E.A. (1983) Solubilization of phospholipids by detergents. Structural and kinetic aspects. Biochim. Biophys. Acta 737:285-304.

Steck, T. (1974) The organization of proteins in the human red blood cell membrane. J. Cell Biol. 62:1-19.

Tanford, C. and Reynolds, J.A. (1976) Characterization of membrane proteins in detergent solutions. Biochim. Biophys. Acta 457:133-170.

Tanford, C. (1978) The hydrophobic effect and the organization of living matter. Science 200:1012-1018.

Tanford, C. (1973) *The Hydrophobic Effect: Formation of Micelles and Biological Membranes.* New York: John Wiley and Sons.

Specific

Ashani, Y. and Catravas, G.N. (1980) Highly reactive impurities in Triton X-100 and Brij 35: Partial characterization and removal. Anal. Biochem. 109:55-62.

Atha, D.H. and Ingham, K.C. (1981) Mechanism of precipitation of proteins by polyethylene glycols. J. Biol. Chem. 256:12108-12117.

Brown, P.J. and Schonbrunn, A. (1993) Affinity purification of a somatostatin receptor-G-protein complex demonstrates specificity in receptor-G-protein coupling. J. Biol. Chem. 268:6668-6673.

Bruns, R.F., Lawson-Wendling, K. and Pugsley, T.A. (1983) A rapid filtration assay for soluble receptors using polyethylenimine-treated filters. Anal. Biochem. 132:74-81.

Cann, J.R. and Hinman, N.D. (1976) Hummel-Dreyer gel chromatographic procedure as applied to ligand-mediated association. Biochem. 15:4614-4622.

Chang, H.W. and Bock, E. (1980) Pitfalls in the use of commercial nonionic detergents for the solubilization of integral membrane proteins: Sulfhydryl oxidizing contaminants and their elimination. Anal. Biochem. 104:112-117.

Colichman, E.L. (1951) Spectral study of long chain quaternary ammonium salts in bromphenol blue solutions. J. Am. Chem Soc. 73:3385-3388.

Edelstein, S.J. and Schachman, E. (1967) The simultaneous determination of partial specific volumes and molecular weights with microgram quantities. J. Biol. Chem. 242:306-311.

Fleming, J.W. and Ross, E.M. (1980) Reconstitution of beta-adrenergic receptors into phospholipid vesicles: Restoration of [^{125}I]-iodohydroxybenzylpindolol binding to digitonin-solubilized receptors. J. Cycl. Nucl. Res. 5:407-419.

Gorrisen, H., Aerts, G., Ilien, B. and Laduron, P. (1981) Solubilization of muscarinic acetylcholine receptors from mammalian brain: An analytical approach. Anal. Biochem. 111:33-41.

Green, A.A. and Hughes, W.L. (1955) Protein fractionation on the basis of solubility in aqueous solutions of salts and organic solvents. Methods in Enzymol. 1:67-90, especially pp. 72-78.

Herberg, J.T., Codina, J., Rich, K.A., Rojas, F.J. and Iyengar, R. (1984) The hepatic glucagon receptor: Solubilization, characterization and development of an affinity adsorption assay for the soluble receptor. J. Biol. Chem. 259:9285-9294.

Hummel, J.P. and Dreyer, W.J. (1962) Measurement of protein-binding phenomena by gel filtration. Biochim. Biophys. Acta 63:530-532.

Jain, M.K. and Wagner, R.C. (1980) *Introduction to Biological Membranes*, pp. 66-70. New York: John Wiley and Sons.

Keith, B., Brown, S. and Srivastava, L.M. (1982) *In vitro* binding of gibberellin A_4 to extracts of cucumber measured by using DEAE-cellulose filters. Proc. Natl. Acad. Sci. USA 79:1515-1519.

Leff, S.E. and Creese, I. (1982) Solubilization of D-2 dopamine receptors from canine caudate: Agonist-occupation stabilizes guanine nucleotide sensitive receptor complexes. Biochem. Biophys. Res. Comm. 108:1150-1157.

Levi, A., Cimino, M., Mercanti, D. and Calissano, P. (1974) Studies on binding of GTP to the microtubule protein. Biochim. Biophys. Acta 365:450-453.

Nexo, E., Hock, R.A. and Hollenberg, M.D. (1979) Lectin-agarose immobilization, a new method for detecting soluble membrane receptors. J. Biol. Chem. 254:8740-8743.

Oka, C., Tsujimoto, R., Kajikawa, M., Koshiba-Takeuchi, K., Ina, J., Yano, M., Tsuchiya, A., Ueta, Y., Soma, A., Kanda, H., Matsumoto, M. and Kawaichi, M. (2004) HtrA1 serine protease inhibits signaling mediated by Tgfβ family proteins. Development 131:1041-1053.

Repaske, M.G., Nunnari, J.M. and Limbird, L.E. (1987) Purification of the α_2-adrenergic receptor from porcine brain using a yohimbine-agarose affinity matrix. J. Biol. Chem. 262:12381-12386.

Reynolds, J.A. and Tanford, C. (1976) Determination of molecular weight of the protein moiety in protein-detergent complexes without direct knowledge of detergent binding. Proc. Natl. Acad. Sci. USA 73:4467-4470.

Rosenthal, K.S. and Koussale, F. (1983) Critical micelle concentration determination of nonionic detergents with Coomassie Brilliant Blue G-250. Anal. Chem. 55:1115-1117.

Rosevear, P., Van Alan, T., Baxter, J. and Ferguson-Miller, S. (1980) Alkyl glucoside detergents: A simpler synthesis and their effects on kinetic and physical properties of cytochrome c oxidase.

Ross, E.M. and Schatz, G. (1976) Cytochrome of bakers' yeast. I. Isolation and properties. J. Biol. Chem. 251:1991-1996.

Schmidt, J. and Raftery, M.A. (1973) A simple assay for the study of solubilized acetylcholine receptors. Anal. Biochem. 52:349-354.

Schneider, W.J., Basu, S.K., McPhaul, M.J., Goldstein, J.L. and Brown, M.S. (1979) Solubilization of the low density lipoprotein receptor. Proc. Natl. Acad. Sci. USA 76:5577-5581.

Smith, S.K. and Limbird, L.E. (1981) Solubilization of human platelet α-adrenergic receptors: Evidence that agonist occupancy of the receptor stabilizes receptor-effector interactions. Proc. Natl. Acad. Sci. USA 78:4026-4030.

Stubbs, G.W., Smith, H.G. Jr. and Litman, B.J. (1976) Alkyl glucosides as effective solubilizing agents for bovine rhodopsin: A comparison with several commonly used detergents. Biochim. Biophys. Acta 426(1):46-56.

Wilson, M.H. and Limbird, L.E. (2000) Mechanisms Regulating the Cell Surface Residence Time of the α_{2A}-Adrenergic Receptor. Biochemistry 39(4):693-700.

6. QUANTIFYING CELL SURFACE RECEPTOR BINDING AND TURNOVER IN INTACT CELLS

The preceding chapters have focused on how to identify and characterize a specific cell surface receptor. Up to this point, the interaction between ligand and receptor has been conceptualized in a rather static framework, as if the receptor molecule remains unchanged in isolated membrane preparations or in detergent-solubilized solution, even after receptor occupancy. This may or may not be true, depending on the properties of the ligand. However, when ligand binding to intact cells is measured, a static conceptualization of the receptor population almost certainly is not valid. Not only is the cell surface membrane a dynamic medium, but receptors continually are being synthesized, internalized, recycled, or degraded, so that the receptor population accessible for interaction with ligand at the cell surface is continually changing. Figure 6-1 provides a schematic diagram of one of the possible fates of a cell surface receptor following ligand occupancy, receptor-mediated endocytosis via clathrin-coated pits. Caveolae represent another morphological subcompartment of the cell surface from which receptors enter the cell. Membrane subcompartments, referred to as lipid rafts, may represent yet another. The present chapter addresses the study of receptors in intact cells and quantitative analyses that can isolate the partial reactions of receptor synthesis, delivery, endocytosis, recycling, and degradation. The text will often refer to the receptor identification strategies as radioligand binding data, but the quantitative methods apply as well to characterization of the fate of ligand and/or receptors using a variety of strategies, as summarized in Table 6-1.

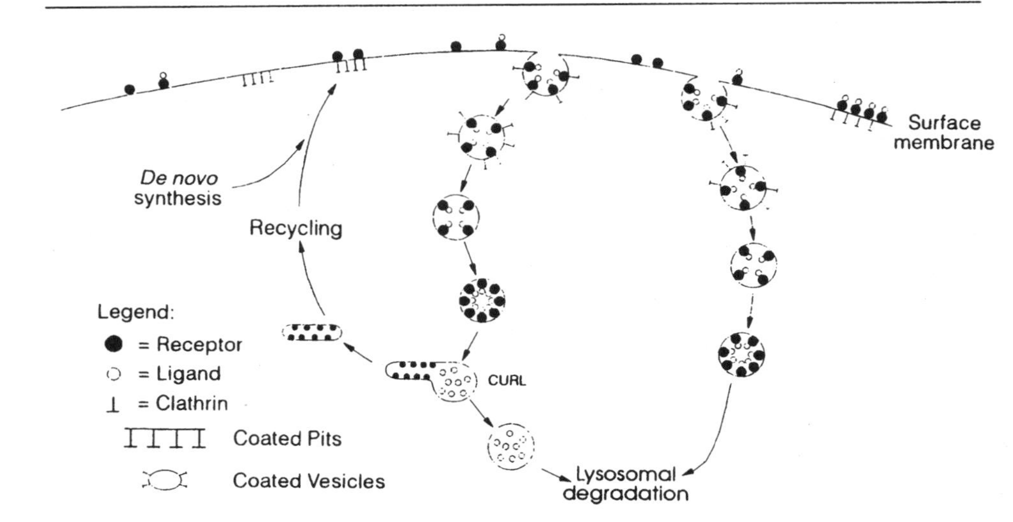

Figure 6-1. The multiple possible fates of a cell surface receptor following ligand occupancy. One "itinerary" for ligand-occupied cell surface receptors includes endocytosis via clathrin-coated pits (Goldstein et al. [1979]; Pearse and Robinson [1990]). Once internalized, these endocytic vesicles shed their clathrin coats. Fusion of these *endosomes* can form a morphologically identifiable tubulo-vesicular **C**ompartment where **U**ncoupling of **R**eceptor from **L**igand (**CURL**) can occur (Geuze et al. [1983]). In many cases, the receptor is spared degradation following only a single round of endocytosis and is returned to the cell surface in a functional state (Brown et al. [1983]). Not all receptor systems recycle, however, and delivery of the receptor to the lysosomes without quantitatively significant recycling probably contributes to the phenomenon of "down regulation," i.e., the decrease in density of functional cell surface receptors occurring after protracted exposure to hormones or agonist drugs (see Carpenter and Cohen [1976] as an example). Clathrin-coated pits are not the only means for receptor removal from the cell surface; caveoli have been postulated to play a role in internalization of G-protein-coupled and other receptor populations (Lisanti et al. [1994]). Recent reviews of cell surface receptor endocytosis provide more detail concerning the molecules and partial reactions involved in these processes (Sorkin and Von Zastrow [2002]; Marchese et al. [2003]).

Many of the morphological sequelae of ligand occupancy of cell surface receptors have been observed to parallel altered biochemical properties of the receptor. This is helpful, because routine monitoring of the fate of cell surface receptors under different experimental conditions often is easier using biochemical strategies rather than exploiting morphological techniques. However, the introduction of strategies to monitor receptor-green fluorescent protein (GFP) fusions also has resulted in semi-quantitative studies of receptor turnover, especially internalization.

DISCRIMINATING BETWEEN CELL SURFACE VERSUS INTRACELLULAR MEMBRANE RECEPTOR-LIGAND COMPLEXES

Accessibility of Impermeant Ligands to Interaction with Receptors

A possible time-dependent fate of a receptor (and ligand) following binding to an intact cell is shown in figure 6-2. The following methods offer independent strategies for quantitating cell surface or internalized ligand. Whereas many of the experimental strategies that permit detection of intracellular receptors depend on prior receptor occupancy by a radioligand as an indicator of receptor distribution, the relative accessibility of hydrophobic versus hydrophilic ligands to receptors in intact cells can provide indirect evidence for whether all or part of the receptor population is accessible at the cell surface, even in the absence of prior receptor occupancy. Because the interior of the biological membrane is apolar, hydrophilic ligands do not equilibrate rapidly across the surface membrane, if they diffuse across at all. In contrast, hydrophobic (lipophilic) ligands penetrate the surface membrane quickly and gain almost immediate access to the cell interior.

Two related experimental approaches for assessing the differential accessibility of hydrophilic versus hydrophobic radioligands to receptors in intact cells is shown schematically in figure 6-2. As shown in figure 6-2A, if a radiolabeled hydrophilic ligand and a radiolabeled hydrophobic ligand both are available, the surface versus inaccessible (presumably internalized) binding can be assessed by comparing the number of binding sites available to the hydrophobic versus the hydrophilic radioligand. The hydrophobic ligand for a particular receptor population would be predicted to have access to all of its unoccupied receptors in the intact cell, i.e., those on the cell surface and those in the cell interior. In contrast, the hydrophilic ligand would be expected to have access solely to receptors on the cell surface, at least during a reasonably short incubation duration. Thus, binding of the radiolabeled hydrophobic ligand would represent the binding of the total receptor population (surface + "internalized") whereas binding of the radiolabeled hydrophilic ligand would identify only those receptors accessible at the cell surface. The difference between the two binding capacities therefore should provide an indirect estimate of the fraction of functional receptors that are in the cell interior.

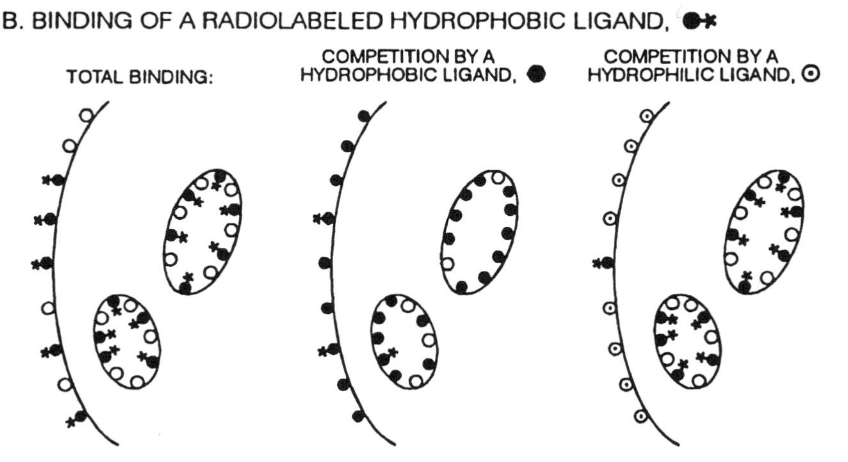

Figure 6-2. Differentiating cell surface from internalized (or inaccessible) receptors using hydrophilic versus hydrophobic ligands.

To prevent receptor redistribution during the incubation intended to assess surface versus internalized receptor location, the binding assays should be performed at 4°C or in the presence of a metabolic inhibitor that prevents receptor redistribution in the intact cell without perturbing ligand binding (Hertel et al. [1985]). If an investigator wants to assess whether agonist or antagonist receptor occupancy influences receptor topography, intact cells can be exposed to these agents for varying periods of time before cooling the cells to 4°C to "freeze" the topographical distribution of receptors and to assay their ability to interact with hydrophobic versus hydrophilic ligands in a radioligand binding assay. Whether unlabeled ligand still is bound to receptors following the pretreatment described above can be determined indirectly by determining the density of receptors detected by a radiolabeled hydrophobic ligand before and after cell pretreatments, since this ligand should have access to all unoccupied receptors and thus could presumably detect loss of detectable receptor binding due to agonist or antagonist pretreatment.

The strategy shown in figure 6-2A was exploited in comparing the binding properties of a radiolabeled hydrophilic antagonist (^{3}H-CGP-12177) with binding properties of several radiolabeled hydrophobic antagonist ligands at the β-adrenergic receptor, including ^{3}H-dihydroalprenolol, ^{3}H-carazolol, and ^{125}I-iodopindolol. Although studies in broken-cell preparations indicated that these radioligands all identify the same total population of β-adrenergic receptors, the binding properties of the radiolabeled (denoted by*) hydrophilic *antagonist ^{3}H-CGP-12177 to *intact* cells differed from those of hydrophobic *antagonist binding. The specific binding of ^{3}H-CGP-12177 to β-adrenergic receptors on intact cells is completely blocked by both agonist and antagonist competitors. Binding of hydrophobic *antagonists is similarly completely blocked by unlabeled hydrophobic antagonists, but is only partially blocked by unlabeled β-adrenergic agonists, typically hydrophilic in structure. Exposing intact cells to unlabeled β-adrenergic agonists followed by extensive washing of the cells prior to radiolabeled binding assays causes a decrease in total binding of the hydrophilic antagonist ^{3}H-CGP-12177, but not in total hydrophobic *antagonist binding. These data are consistent with the interpretation that pretreating intact cells with β-adrenergic agonists causes a redistribution of β-adrenergic receptors so that they become inaccessible to hydrophilic ligands (Staehelin et al. [1982]; Staehelin and Simons [1982]; Staehelin and Hertel [1983]; Kallal et al. [1998]).

As shown in figure 6-2B, if a hydrophilic ligand for the receptor is available in unlabeled form, the accessibility of receptors for binding to this agent can be determined, indirectly, by the ability of this hydrophilic ligand to block binding of a radiolabeled hydrophobic ligand to the total cellular receptor population. The hydrophobic radioligand would be expected to identify all functional cell-associated receptors-those at the cell surface as well as those in the cell interior. In contrast, the hydrophilic ligand would be expected to block only that component of hydrophobic radioligand binding which represents receptor binding at the cell surface. These comparisons of competition for the binding of a radiolabeled hydrophobic ligand by hydrophilic versus hydrophobic competitors often are evaluated in very short-duration pre-equilibrium incubations, since under true equilibrium binding conditions, the radioligand and competitors all will have been allowed to reach their own equilibrium. It has been observed in a number of model systems that agonist occupancy of β-adrenergic receptors in intact cells results in a transient high affinity receptor-agonist interaction that is quickly (seconds to minutes) followed by a decrease in apparent receptor affinity for agonists, as assessed by the competition of unlabeled hydrophilic agonists for hydrophobic *antagonist binding. Exposing cells to antagonist agents does not cause this phenomenon, and any change in EC_{50} values for antagonists as a function of incubation duration can be accounted for entirely by changes expected for equilibration of radioligand and competitor when both are interacting with a receptor population via mass action law (Pittman and Molinoff [1980]; Insel et al. [1983]; Toews and Perkins [1984]). The transient nature of high-affinity receptor-agonist interactions noted in intact cell binding assays may reflect the speed with which agonists provoke a redistribution of the receptor population to a compartment not readily accessible to hydrophilic ligands.

The observation that a *hydrophilic antagonist* (sotalol) also demonstrates apparent high- and low-affinity interactions with β-adrenergic receptors in intact cell radioligand binding assays after cells are exposed to agonist (but not antagonist) agents has suggested the possibility that time-dependent changes in *apparent* receptor affinity for agonists might be a manifestation of internalization or sequestration of receptors to a compartment inaccessible to hydrophilic ligands. Sotalol, like the hydrophilic agonist isoproterenol, attains equilibrium very rapidly with the apparent high-affinity receptors but equilibrates only slowly with the fraction of the β-adrenergic receptor population that exhibited lower apparent affinity in short-term (i.e., preequilibrium) assays. These data have been interpreted to indicate that pretreating cells with an agonist provokes a redistribution of a fraction of the receptor population to an internalized, or inaccessible, pool that demonstrates a low apparent affinity for hydrophilic ligands in short-time assays. The subsequent slow equilibration of hydrophilic ligands with these lower affinity receptors is really a manifestation of time-dependent accessibility of the hydrophilic ligand to this sequestered receptor population provoked by agonist occupancy of the receptor (Toews and Perkins [1984]; Toews et al. [1984]).

Examining the change in apparent affinity of hydrophilic ligands in competing for a hydrophobic competitor can assess the ability of a particular receptor structure to redistribute following agonist occupancy (Parker et al. [1995]). This strategy has been used to compare the ability of recombinant α_2-adrenergic receptor subtypes (α_{2A}, α_{2B}, and α_{2C}) expressed in heterologous cells to be removed from the cell surface in response to agonist occupancy (Edson and Liggett [1992]; Picus et al. [1993]).

Protease-Resistant Ligand Binding as a Measure of Internalized Receptor-Ligand Complexes

Proteases can rapidly hydrolyze polypeptide hormones. This property has been exploited to ascertain whether cell-associated radioactivity is due to surface-bound (protease-accessible) or internalized (protease-inaccessible) binding of radioactively labeled peptide hormones. Using protease treatment to remove ligand-receptor complexes accessible at the cell surface is applied most frequently to polypeptide hormone binding to cultured cells. However, proteolytic treatment probably removes ligand-receptor complexes at the cell surface not only by digesting the radiolabeled polypeptide hormone but also by hydrolyzing the exofacial domain of the receptor. Therefore, this method may be suitable to remove binding due to cell surface ligand-receptor complexes even in situations where the radioligand is not labile to proteolytic hydrolysis (Simantov and Sachs [1973]). Fig. 6-3 provides a schematic of how discriminating surface from intracellular ligand can help describe the fate of the receptor and the ligand following initial receptor occupancy.

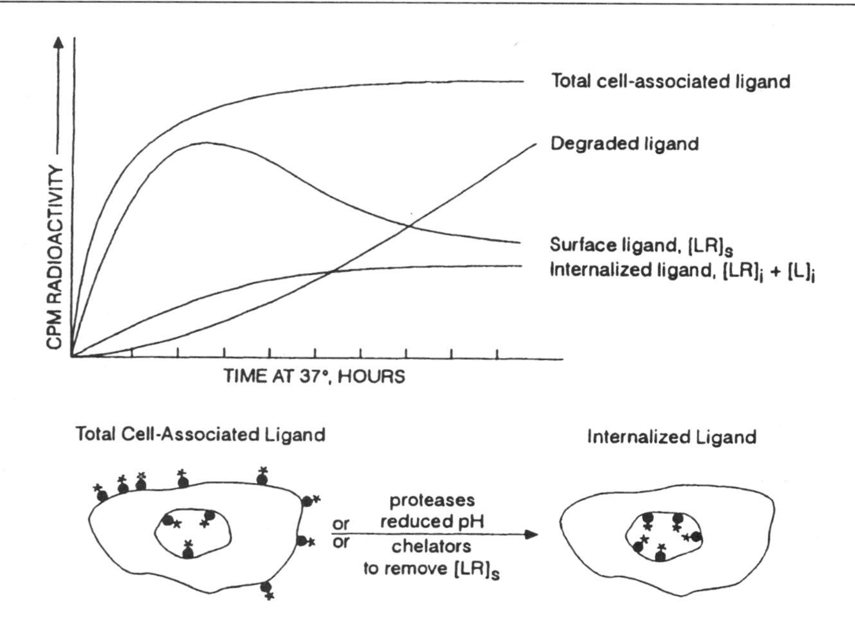

Figure 6-3. Radioligand binding to intact cells: assessment of the surface versus intracellular distribution of total cell-associated radioactivity as a function of time. Using impermeant ligands, protease removal of surface-associated ligand or receptor, or rapid dissociation of surface-bound ligand secondary to reduced pH or other manipulations, a progress curve can be determined which charts the fate of surface-bound ligand to the cell interior.

In these experiments, the radiolabeled hormone is incubated with intact cells for varying periods of time at 37°C. At selected intervals, further turnover of the receptor population is blocked by cooling the cells to 4°C or by adding metabolic inhibitors to block endocytosis, and unbound ligand is removed by washing the cells with ice-cold isotonic buffer. At this point, the ligand's fate can be resolved into three categories: (1) surface-bound, (2) internalized (free or receptor-bound), or (3 internalized, degraded, and subsequently released. The total amount of cell-associated radioactivity is then determined on some of the cell preparations to yield an estimate of the quantity of surface-bound plus internalized ligand. Other aliquots (or other culture wells) of the washed, chilled cells are incubated with proteases under conditions previously determined to permit removal of radioligand bound to cell surface receptors. (The conditions for protease digestion are optimized by using cells that have been incubated under conditions-e.g., 4°C-where all specifically bound hormone should represent surface-bound ligand.) The amount of specific ligand binding that remains cell-associated following the protease treatment is a measure of internalized ligand, whereas the quantity of cell-associated ligand due to binding by receptors at the cell surface is that binding which is lost upon protease digestion.

The cell-associated radioactivity that remains following protease digestion can be due to (1) internalized receptor-ligand complexes, or (2) internalized ligand that has dissociated from the receptor. Dissociated ligand may represent native hormone or hormone that has been hydrolyzed. To resolve receptor-bound ligand from dissociated ligand and ligand fragments, protease-treated cells can be solubilized by a suitable detergent (e.g., Triton X-100) and receptor-bound ligand can be isolated from the detergent extract by precipitation or immunoisolation (cf. chapter 5). Determination of whether the internalized ligand *not* bound to receptors is native hormone or has been hydrolyzed to a small molecular weight species can be assessed by HPLC or SDS-polyacrylamide electrophoresis. In the latter case, detection of the hormone and/or its fragments is subsequently achieved by autoradiography; which is most easily accomplished for radioiodinated peptide and protein ligands (see Stoscheck and Carpenter [1984]).

Rapid Dissociation of Exofacial Ligand-Receptor Complexes to Resolve Surface-bound from Internalized Ligand

It has been observed in several ligand-receptor systems that surface-bound radioligand can be dissociated by lowering extracellular pH (Haigler et al. [1980]). In these situations, the fraction of total cell-associated binding removed by acid exposure of cells is interpreted to represent surface-accessible binding. To prevent receptor population turnover and/or degradation of receptors during exposure to reduced pH (i.e., pH 3.0-4.5), cells are cooled to 4°C before treatment with acid. Two criteria are used to optimize procedures for dissociation of hormone-receptor complexes at reduced pH. First, dissociation conditions are selected that remove all cell-associated radioligand bound during incubation conditions selected to ensure that all detected binding reflects cell-surface binding (e.g., binding at 4°C or in the presence of sodium azide or phenlyarsine oxide, i.e. agents that inhibit cellular metabolism and thus disrupt ATP-dependent processes such as receptor-mediated endocytosis). Second, the viability of the cellular receptor system after exposure to reduced pH is assessed by determining the ability of radioligand to rebind to the dissociated receptors and the receptor's ability to elicit its characteristic response in the cell. Optimal conditions are those that permit dissociation of all surface-bound ligand without inhibiting subsequent receptor-ligand interactions and receptor-mediated response. The validity of using reduced pH to dissociate ligand from receptors at the cell surface and thus distinguish surface-bound from internalized radioligand can be documented further by performing correlative studies, e.g., by using quantitative electron microscopic autoradiography (Carpenter and Cohen [1976]).

In some circumstances, reducing extracellular pH may not perturb the electrostatic interactions between ligand and receptor at the cell surface. For example, LDL interactions with cell surface LDL receptors are not dissociated by decreasing the pH in the medium, but can be dissociated quantitatively by adding heparin, suramin or other glycosaminoglycans to the incubation (Goldstein et al. [1976]). Similarly, adding the divalent metal cation chelator EDTA to the incubation medium terminates the interaction of asialoproteins with the asialoglycoprotein receptor. Despite the particular manipulation used to release surface-bound radioactivity, the same rationale applies as for acid-labile interactions, as do the same controls for assessing the validity of the biochemical approach for quantitatively removing surface-bound ligand without perturbing the functional properties of the receptor.

BIOCHEMICAL EVIDENCE CONSISTENT WITH RECYCLING OF CELL SURFACE RECEPTORS

A number of cell surface receptor populations have been observed to recycle to the cell surface following internalization of ligand-receptor complexes. Evidence for receptor recycling often can be obtained by comparing the number of ligand molecules internalized per cell with the number of ligand-binding receptors per cell, particularly when *de novo* receptor synthesis is demonstrated to be negligible during the time course over which ligand uptake is measured. If the number of ligand molecules accumulated per cell by a receptor-specific mechanism far exceeds the number of receptors per cell, the existence of receptor recycling is suggested.

Four general phases exist in any experimental design to evaluate receptor recycling: (1) determining the length of time it takes for one "round" of internalization to occur; (2) permitting one round of internalization to occur, followed by inactivation of all residual cell surface receptors by protease treatment; (3) permitting the reappearance of internalized receptors, if it occurs; and (4) determining whether receptors that subsequently appear on the cell surface represent new or recycled receptors.

One approach for determining the length of time it takes to complete one round of intenalization involves binding radioligand to intact cells incubated at 4°C under conditions where only receptors expressed at the cell surface will be able to interact with the ligand. After removal of free ligand by washing at 4°C, the amount of binding attained is measured. Separate control experiments demonstrating that all of the binding is susceptible to proteolytic digestion or is removed by reduced pH, etc., can document that all binding observed during this incubation is indeed to receptors at the cell surface. To determine how long it takes for the cell surface receptor-ligand complexes to be internalized, labeled and washed, cells are warmed to 37°C and the rate and extent of internalization as a function of time of incubation at 37°C is monitored by assessing the accessibility of the ligand to proteases or other perturbations known to discriminate surface-bound from internalized ligand.

Once the time required for one round of internalization has been determined, the possible recycling of cell surface receptors can be addressed. In these studies, cell surface receptors are occupied by ligand in a 4°C incubation or left unoccupied prior to warming the cells to 37°C to permit one round of internalization to occur. Any residual cell surface receptors are then removed from the surface by chilling the cells to 4°C and inactivating the cell surface binding sites, e.g. by proteolysis. After appropriate washing of the cells, recovery of receptors on the cell surface is monitored as a function of time at 37°C. The appearance of functional receptors on the cell surface at this point can be attributed to recycling of previously internalized receptors or to insertion of newly synthesized receptors into the surface membrane. Protein synthesis inhibitors could block appearance of newly synthesized receptors, but findings may be confounded by the known role of rapidly turning over proteins in vesicular trafficking of cell surface proteins (Krupp and Lane [1982]; Krupp et al. [1982]).

A second approach for direct biochemical assessment of receptor recycling relies on the availability of an irreversible ligand for the receptor. For example, cell surface receptors for insulin on rat hepatocytes have been covalently labeled with the biologically active photoprobe ^{125}I-[2-nitro-4-

azidophenylacetyl^{B2}] des-Phe^{B1}-insulin and then permitted to internalize, as assessed by a time-dependent loss of trypsin sensitivity of the radiolabeled binding. Subsequently, there is a progressive reappearance of ligand-receptor complexes at the cell surface, as indicated by the recovery of trypsin sensitivity of the labeled insulin receptors. The interpretation of these findings as evidence for receptor recycling has been documented by quantitative electron microscopic autoradiography performed on parallel preparations (Fehlmann et al. [1982]).

Reversible biotinylation strategies represent a third biochemical approach to evaluate and quantify receptor recycling. In these experiments, receptors at the cell surface (in fact, all molecules on the cell surface) are covalently modified at 4°C by Sulfo-NHS-Biotin (Pierce), a bifunctional biotinylation reagent which is cleavable with disulfide reagents such as dithiothreitol (DTT) or MESNA. The time course of receptor internalization is first assessed. The fraction of DTT-accessible biotinylated receptors at the surface at time 0 is defined as the cell surface receptor population. Cells are then warmed to 37°C and the time-dependent distribution of receptors to an intracellular compartment (+/- hormone or receptor agonist) is allowed to occur. Each time point is terminated by incubation on ice and treatment with DTT to remove surface-accessible receptors, as DTT cleaves the biotinylating reagent accessible to the extracellular compartment. As receptors internalize, a larger fraction of the total cell surface population "marked" by biotinylation at time 0 becomes resistant to DTT treatment, because the endocytosed receptor is now in a DTT-inaccessible compartment. To quantify biotinylated receptors, cells treated at each time point are extracted into biological detergent, biotinylated proteins are isolated using streptavidin-agarose, and the receptor is detected (quantitated) by Western analysis, using antibody against native receptor or a commercial antibody against an epitope-tagged receptor introduced into heterologous cells using cDNA expression strategies.

Once an optimal time point for detecting internalized (DTT-inaccessible, biotinylated) receptor is ascertained, the process of receptor recycling can be evaluated. Cell surface proteins are again biotinylated at 4°C at time 0, and treatment (or not) with a receptor ligand occurs for a specified period of time previously determined to allow accumulation of a detectable, internalized pool of receptor. Now, all cell surface biotin is released by exposure, at 4°C, of the cells to DTT or other reducing agents. The cells are warmed again, and time-dependent reappearance of biotinylated receptors to the cell surface (rendering their biotinylation DTT-sensitive) can be assessed (cf. Turvy and Blum [1998]).

Table 6-1 Non-radioligand strategies for quantitative assessment of receptor redistribution from (and in some cases to) the cell surface.

Strategy	Receptor	Partial Reactions Assessed*	Reference(s)
Reversible Biotinylation	transferrin receptor; MHC molecules	E,R	Turvy and Blum [1998]
Biotinylated Ligand	transferrin receptor	E,R	Vieira [1998]
Cell Surface ELISA	epitope-tagged α_2-adrenergic receptors	E	Daunt et al. [1997]
Confocal Microscopy	native neurokinin receptors in enteric neurons	E	Jenkinson et al. [1999]
	GFP-tagged GPCR (PTH receptor)	E	Sneddon et al. [2003]; Kallal et al. [1998]
	epitope-tagged GPCR (5HT2A receptor)		Bhatnagar et al. [2001]
Epifluorescence Microscopy (fluorescence ratio for quantitation)	opioid receptors	E,R	Tanowitz and von Zastrow [2003]
FACS (Fluorescence-Activated Cell Sorting)	β_2-adrenergic receptors	E	Barak [1994]
	CRLR/RAMP1	E	Hilairet et al. [2001]

* E = endocytosis; R = recycling

A variety of biochemical and quantitative morphological strategies have been developed or adapted for assessment of cell surface receptor turnover in addition to the radioligand binding or reversible biotinylation strategies outlined above, examples of which are summarized in Table 6-1. What follows is a discussion of how to estimate rate constants for receptor delivery, endocytosis and recycling of receptors based on the experimental data that can be obtained using any of these strategies.

ASSESSMENT OF RATE CONSTANTS FOR RECEPTOR TURNOVER USING A STEADY STATE MATHEMATICAL ANALYSIS OF INTACT CELL RADIOLIGAND BINDING DATA

The foregoing discussions of experimental approaches describing the fate of ligand-occupied cell surface receptors have emphasized that cell surface receptors are being continually synthesized, internalized, perhaps recycled, and degraded. Binding to intact cells under physiological conditions cannot be analyzed using *equilibrium* mathematical models such as Scatchard analysis, because receptors on cells incubated above 4°C are *not* in equilibrium with their environment.

A *steady state* model for analyzing cellular binding, internalization, and degradation of polypeptide ligands was first introduced by Wiley and Cunningham in 1982. This mathematical approach is useful for determining the rate of clearance of ligand-receptor complexes from the cell surface as well as the rate of insertion of new receptors into the surface. Consequently, phenomena such as receptor down-regulation, receptor induction, or other changes in the steady state concentration of cell surface receptors can be described quantitatively in terms of altered rates of receptor clearance, insertion, or both. The mathematical descriptions that comprise this steady state approach are summarized below. Although the methodological approach and data analysis are described as if receptor identification and turnover are measured using radioligand binding, data obtained using other strategies are equally amenable to these analyses. Extensions of the model also permit assessments of recycling, or ligand-dependent changes in sorting post-endocytosis (cf. Lauffenburger and Linderman [1993]).

If the functional cell surface receptors are at steady state with the total cellular receptor population, then the rate of entry into the cell surface will be equivalent to the rate of removal from the cell surface, or

$$V_r = k_i[R]_s \tag{6.1}$$

where V_r = rate of insertion of new receptors into the cell surface
$[R]_s$ = surface concentration of unoccupied receptors
k_t = rate constant for turnover of unoccupied receptors, min^{-1}

Ligand binding to the receptor may or may not change the rate at which receptors are internalized. However, once a new steady state is achieved following ligand occupancy, the surface concentration of occupied and unoccupied receptors, by definition, will be a constant, such that

$$V_r = k_t[R]_s + k_e[LR]_s \tag{6.2}$$

where $[LR]_s$ = surface concentration of ligand-receptor complexes
k_e = endocytotic rate constant, min^{-1}, of occupied receptors

Equation 6.2 can be rearranged to give

$$[R]_s = \frac{V_r - k_e[LR]_s}{k_t} \tag{6.3}$$

At steady state, the rate of formation of ligand-receptor complex will equal its rate of loss by dissociation plus its rate of loss by internalization, such that

$$k_a [L][R]_s = k_d[LR]_s + k_e [LR]_s \tag{6.4}$$

where k_a = rate constant for association of the ligand-receptor complex, in units of M^{-1}, min^{-1}
k_d = rate constant for dissociation of the ligand-receptor complex, min^{-1}
$[L]$ = concentration of free ligand in the incubation medium, M

Rearrangement of equation 6.4 provides an expression for the concentration of occupied receptors at the cell surface:

$$[LR]_s = \frac{k_a[L][R]_s}{k_d + k_e} \tag{6.5}$$

and dividing both sides by $[L]$ yields:

$$\frac{[LR]_s}{[L]} = \frac{k_a[R]_s}{k_d + k_e} \tag{6.6}$$

Substituting into equation 6.6 with the expression for $[R]_s$ provided in equation 6.3 yields:

$$\frac{[LR]_s}{[L]} = -\left[\frac{k_e k_a}{k_t(k_d + k_e)}\right][LR]_s + \frac{V_r k_a}{k_t(k_d + k_e)} \tag{6.7}$$

Equation 6.7 is an equation of a straight line ($y = mx + b$).

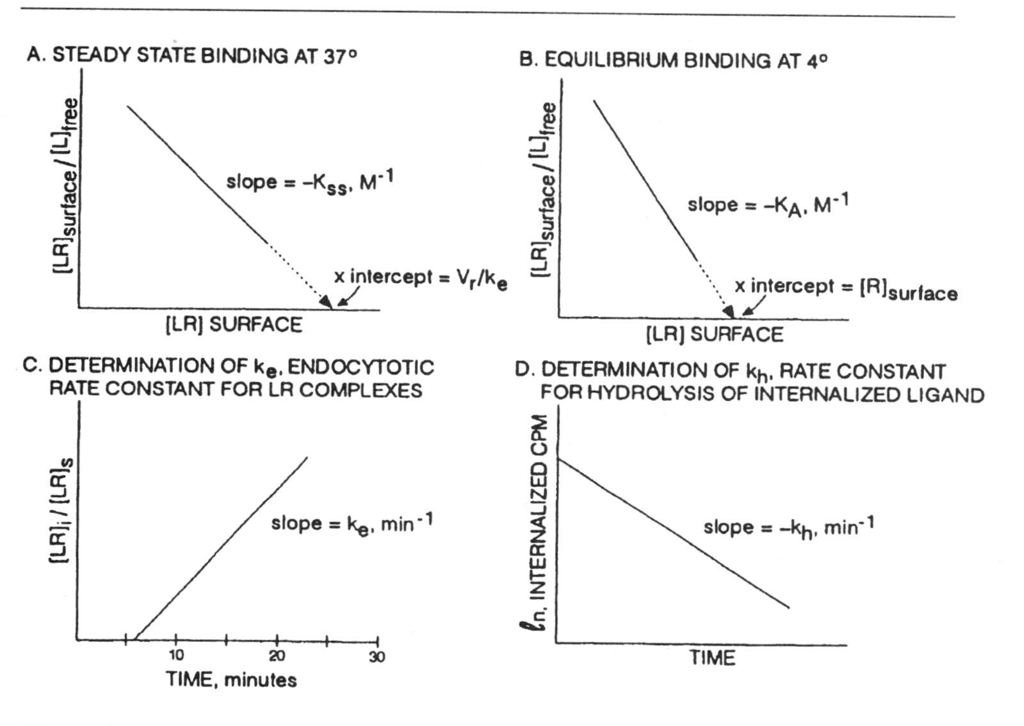

Figure 6-4. Data transformations useful for obtaining parameters for receptor turnover using steady state mathematical analysis of intact cell binding.

As shown in figure 6-4, a plot of $[LR]_s/[L]$ versus $[LR]_s$ should yield a straight line with a slope of $-k_e k_a/k_t(k_d + k_e)$. The x intercept (i.e., when $y = 0$) will be V_r/k_e. The "lumped" constant that comprises the slope of this line has been designated K_{ss} (steady state constant) by Wiley and Cunningham, such that

$$K_{ss} = \frac{k_e k_a}{k_t(k_d + k_e)}, \text{ units} = \text{M}^{-1}$$

Plotting the data according to equation 6.7 resembles the data transformation for constructing a Scatchard plot. However, the interpretations of the slopes and intercept are different because equilibrium conditions did not necessarily prevail during the experimental incubation. Consequently, K_{ss} is not a valid measure of receptor affinity for ligand because internalization and other aspects of receptor turnover may be ongoing during an intact cell binding experiment. Likewise, the x intercept is not a manifestation of the total number of cell surface receptors, but is equal to V_r/k_e. However, it is worth noting that if ligand occupancy of cell surface receptors does not alter the rate of receptor internalization, then $k_e = k_t$, which means that the x intercept (V_r/k_e) also equals V_r/k_t, which, as shown in equation 6.1, equals $[R]_s$, the total number of cell surface receptors.

The amount of ligand bound to intact cells frequently represents the sum of surface-bound ligand plus internalized ligand. Consequently, an equation has been derived for *total* cell-associated ligand that is analogous to equation 6.7. If measurements are made when the pool of intact ligand in the cell has reached steady state, then the rate of ligand entry into this pool by internalization equals the rate of loss by degradation, such that:

$$k_e\,[LR]_s = k_h\,[LR]_i \tag{6.8}$$

where $[LR]_I$ = concentration of ligand-receptor complex inside the cell
k_h = rate constant for ligand hydrolysis inside the cell, min^{-1}

The assumption in equation 6.8 that both ligand internalization and ligand degradation are first-order processes has been demonstrated in several model systems (see Wiley and Cunningham [1982]). Rearranging equation 6.8 yields an expression for internalized ligand-receptor complexes:

$$[LR]_i = \frac{k_e[LR]_s}{k_h} \tag{6.9}$$

If the total amount of cell-associated ligand is receptor-bound ligand, then:

$$[LR]_T = [LR]_s + [LR]_i \tag{6.10}$$

combining equations 6.10 and 6.9,

$$[LR]_T = [LR]_s\left(1+\frac{k_e}{k_h}\right) \tag{6.11}$$

Since $[LR]_s$ can be related to $[LR]_T$ as shown in equation 6.11, then the equation for a straight line given in equation 6.7 for $[LR]_s/[L]$ can be converted to:

$$\frac{[LR]_T}{[L]} = -K_{ss}[LR]_T + \left[\frac{V_r k_a}{k_t\left(k_d + k_e\right)}\right](1 + k_e/k_h) \tag{6.12}$$

Again, this equation describes a straight line with a slope of $-K_{ss}$ and an intercept at the x axis of $V_r/k_e(1 + k_e/k_h)$.

The relationships defined above permit experimental determination of several rate parameters for cell surface receptor turnover. For example, the rate of endocytosis of ligand receptor complexes (k_e) can be determined in the following way. If it is assumed that there is no degradation of the internalized ligand during the course of the experiment, then it follows that at steady state the change in concentration of ligand-receptor complexes in the interior will be entirely a function of the rate of internalization of ligand-receptor complexes from the cell surface, or:

$$\frac{d[LR]_i}{dt} = K_e[LR]_s \quad (6.13)$$

Rearranging and integrating the above equation yields the following equation for a straight line:

$$\frac{[LR]_i}{[LR]_s} = k_e t \quad (6.14)$$

where t = time after addition of the ligand. A plot of $[LR]_i/[LR]_s$ versus time t will yield a slope of k_e. This linear relationship, however, requires that (1) there is no degradation of the internalized ligand, and (2) $[LR]_s$ remains constant, i.e., the system is truly at steady state.

However, as stated earlier (equation 6.8), if measurements are made when the pool inside the cell has reached a steady state, then

$$k_e[LR]_s = k_h[LR]_i$$

and this equation can be rearranged to:

$$\frac{[LR]_i}{[LR]_s} = \frac{k_e}{k_h} \quad (6.15)$$

This equation can be used to determine relative internalization and degradation rate constants, k_e and k_h, respectively.

Finally, the half-life ($t_{1/2}$) of an internalized ligand is directly related to the first-order rate constant of ligand hydrolysis inside the cell:

$$k_h = \frac{0.693}{t_{1/2}} \quad (6.16)$$

This relationship can be used for directly determining the value of k_h in an experimental system.

To use the above steady state equations for assessment of receptor turnover in intact cells, the assumptions inherent in the derivation need to be met in the experimental system being evaluated:

1. The experimental system under study must actually reach a steady state with respect to binding, internalization, and degradation of ligand. To determine the time to reach steady state, plots of $[LR]_i$ and $[LR]_s$ versus time are constructed. When steady state is attained, there will be no further change in $[LR]_i$ or $[LR]_s$ versus time, i.e., each parameter will reach a "plateau."

Wiley and Cunningham (1982) demonstrated that the endocytotic rate constant for ligand-receptor complexes (k_e) can be determined before steady state if $d[LR]_s/dt$ is *approximately* zero (at true steady state, $d[LR]_s/dt$ equals zero). This approximation usually can be attained if the period of measurement used to determine k_e is sufficiently short and if $d[LR]_i/dt >>> d[LR]_s/dt$.

2. It is essential that the values for $[LR]_s$ and $[LR]_i$ are determined with great accuracy, which requires that the quantitative limits of the method used to remove surface-bound ligand are rigorously determined. Several experimental approaches for removing surface-bound ligand were described earlier (proteases, low pH, chelators, etc.). The effectiveness of such procedures is estimated by determining the amount of cell-associated radioactivity that remains after removing surface-bound radioactivity when binding to the cells is performed under experimental conditions that prevent receptor internalization, e.g., incubation at 4°C or in the presence of a metabolic inhibitor. If it is determined that the method used for removing cell surface ligand removes 90% (but not 100%) of the ligand, then the value of $[LR]_s$ estimated experimentally must be corrected before inserting a value for $[LR]_s$ into the steady state analysis. Often, methods for removing cell-surface ligand also remove a small fraction of internalized ligand, which also can be quantitated for each experimental system (see Wiley and Cunningham [1982]). If it is determined that the method used for removing cell-surface ligand removes 3% of internalized ligand, for example, then the values of $[LR]_i$ obtained experimentally must be corrected before substituting them into the data transformations used for steady state analysis. This attempt to accurately determine the values of $[LR]_s$ and

$[LR]_i$ is not undertaken to permit a calculation of the endocytotic rate constant k_e to the third decimal point; rather, it is related to eliminating all technical error so that deviations from predicted behavior can be interpreted in biological terms. For example, if a plot of $[LR]_s/[LR]_i$ versus t is nonlinear, then it can be assumed that some of the assumptions made for determining k_e (equal to the slope of the above plot) were not met experimentally, e.g., ligand was degraded during the course of the experiment. A change in the slope of a plot of $[LR]_s/[LR]_i$ versus t usually indicates the time at which ligand degradation starts to occur.

3. Since the determination of k_e by the steady state approach derived by Wiley and Cunningham requires that no ligand degradation occur during the time course of the experiment, it is necessary to determine an interval of time during which no ligand degradation takes place experimentally. This can be accomplished by incubating cells with ligand at 4°C, warming the cells to 37°C to permit internalization, removing cell-surface ligand at various time points, and assessing degradation of internalized ligand by appropriate analytical methods, such as HPLC or SDS-polyacrylamide gel electrophoresis.

4. The concentration of ligand used in the steady state plots is the concentration of *free* ligand, not the concentration of ligand added to the incubation. For polypeptide hormones, the concentration of free radioligand available can be determined by TCA precipitation of the incubation medium upon termination of the incubation. The TCA-precipitable counts are assumed to represent intact hormone, and this assumption can be documented by appropriate analysis of the TCA-precipitated material.

Once it has been documented that the assumptions inherent in the steady state analysis are met in the experimental system of interest, several parameters of receptor turnover can be determined using steady state analysis:

k_e, min^{-1}	rate constant for endocytosis of ligand-receptor complexes
k_t, min^{-1}	rate constant for endocytosis of unoccupied receptors
k_h, min^{-1}	rate constant for degradation of internalized ligand
V_r	rate of insertion of new receptors (units of receptors/cell/min)
K_{ss}, M^{-1}	steady state constant

As shown in figure 6-4A, a plot of $[LR]_s/[L]$ versus $[LR]_s$ yields a straight line whose slope equals $-K_{ss}$ and whose intercept is V_r/k_e. If steady state has been attained, then a plot of $[LR]_T/[L]$ also should yield a straight line, and the slope ($-K_{ss}$) should be the same value as that obtained when the binding data plotted represent surface-bound ligand (i.e., $[LR]_s$) only.

When plots such as those in figure 6-4A are constructed at 4°C rather than at 37°C and the incubation duration is sufficient to attain equilibrium, then one is performing a Scatchard analysis and the intercept on the *x* axis for cell-associated radioactivity equals $[R]_s$, the concentration of ligand-receptor complexes on the cell surface. The slope of this equilibrium plot of $[LR]_s/[L]$ versus $[LR]$ is $-K_A$, the equilibrium association constant for interaction of ligand with receptor, in units of M^{-1}. The value of $[R]_s$ determined from this plot is of value, used in equations for determining k_t, the rate constant for turnover of unoccupied receptors (see later).

The rate constant for internalization of ligand-receptor complexes (k_e) can be determined in the following manner. As shown in figure 6-4C, a plot of $[LR_i]$ $[LR]_s$ versus time should yield a straight line with a slope of k_e, min^{-1}. As mentioned previously, divergence from a straight line usually indicates that degradation of internalized ligand has taken place during the measurement. Since determination of k_e requires that ligand degradation does not occur during the interval when $[LR]_s$ and $[LR]_i$ are measured for estimation of k_e, it is most convenient in these experiments to bring cells to steady state with unlabeled ligand. When steady state occupancy is attained, the cells are chilled to 4°C and quickly but extensively washed with ice-cold buffer to remove unlabeled ligand. The cells are then warmed to 37°C in the presence of radiolabeled ligand, and $[LR]_s$ and $[LR]_i$ are determined at varying time points over an interval of time where degradation of the internalized radioligand previously has been documented not to occur (Waters et al. [1990]).

To determine the value for k_h, the rate of degradation of internalized ligand, the ratio of $[LR]_i/[LR]_s$ is determined at steady state, i.e., when $[LR]_i$ and $[LR]_s$, no longer change as a function of time. From equation 6.15, the ratio of $[LR]_i/[LR]_s$ can be shown to equal k_e/k_h. Consequently, k_h can be obtained by substituting the observed ratio of $[LR]_i/[LR]_s$ determined at steady state and the value of k_e determined above into equation 6.15. The internal consistency of the data can be assessed by determining this parameter directly by bringing to steady state with radioligand, then switching to a medium containing unlabeled hormone. A plot of *in* (internalized ligand, cpm) versus time should yield a straight line, the slope of which is equal to $-k_h$ (figure 6-4D). If the data are plotted as $\log_{10}$ (internalized ligand, cpm) versus time, the slope will equal $-2.303\ k_h$.

The value of V_r, the rate constant for insertion of new receptors into the cell surface, also can be obtained from steady state analyses of intact cell binding. From equation 6.7 it can be shown that a plot of $[LR]_s/[L]$ versus $[LR]_s$ yields a straight line whose *x* intercept is V_r/k_e (cf. figure 6-4A).

Multiplying the intercept of this plot times k_e (as determined above) yields the value of V_r.

The fourth rate parameter that is obtained conveniently from steady state analysis is that of k_t, the rate constant for turnover of unoccupied receptors. From equation 6.3 it can be shown that $[R]_s = V_r - k_e[LR]_s/k_t$. Values of V_r and k_e for substitution into this equation can be obtained as described above. To obtain values for $[R]_s$ and $[LR]_s$ to substitute into equation 6.3, a set of cells is brought to steady state with a subsaturating ligand concentration designated $[L]^x$ (where x refers to a specific concentration). After the amount of ligand bound to the cell surface is measured at steady state (to determine $[LR]^x$), the rest of the set of cells is rapidly cooled to 4°C (to prevent cell-surface receptor turnover) and the total number of cell-surface receptors obtained by performing equilibrium Scatchard analysis as in figure 6-4B. The x intercept of this plot is the *total* density of cell surface receptors, or $[LR]^x + [R]^x$. The value for k_t is obtained from the equation

$$k_t = \frac{V_r - k_e[LR]^x}{[R]^x}$$

The advantage of the steady state analysis described above is that several rate parameters for receptor turnover can be obtained without the need to add reagents, such as cycloheximide (Kadle et al. [1983]), which could influence multiple cellular processes. For example, this approach can assess whether down-regulation of cell surface receptors induced by heterologous versus homologous agents results from an accelerated rate of endocytosis or a reduced rate of insertion, or both (for an example, see Lloyd and Ascoli [1983]). Similarly, comparing the calculated values for k_e versus k_t can determine whether ligand occupancy accelerates receptor internalization (Ronnet et al. [1983]).

The steady state approach introduced by Wiley and Cunningham has been modified by others (e.g., Schwartz et al. [1982]) to include certain simplifying assumptions that have been tested for their validity in the experimental systems where employed. In addition, this conceptual approach has been extended to *in vivo* studies of receptor turnover and has been successfully applied to the description of changes in β-adrenergic receptor turnover following agonist infusions in whole rats (Snavely et al. [1985]).

The calculations outlined above provide numerical descriptors for the insertion, occupancy and removal of receptors from the cell surface. Molecular details concerning the various topological itineraries for cell surface receptors continue to emerge, and mathematical models that further resolve the endocytotic process have been developed to quantify newly revealed partial reactions (Wiley [1992]). Quantitative descriptors of the endocytosis of EGF-EGF receptor complexes via smooth versus clathrin-coated pits, and receptor recycling versus independent ligand and receptor degradation, have been modeled (Starbuck and Lauffenburger [1992]), as has regulation of cell surface receptor density via endosomal retention (French et al. [1994]; Herbst et al. [1994]).

THE HEAVY AMINO ACID DENSITY-SHIFT TECHNIQUE FOR QUANTITATING RECEPTOR SYNTHESIS AND TURNOVER

Devreotes and associates (1977) introduced an ingenious method for documenting the rate of appearance of newly synthesized cell surface receptors in the study of acetylcholine receptor turnover in denervated skeletal muscle. The elegance of this method is that receptor synthesis can be followed even when antireceptor antibodies or techniques for receptor purification are not available. The only requirement is that a living (usually cultured) cell system be available so that *de novo* receptor synthesis can be monitored.

The rationale behind this approach is that substituting heavy, isotope-labeled amino acids (^{15}N, ^{3}C, ^{2}H) for normal, light amino acids (^{17}N, ^{12}C, ^{1}H) in the culture medium of the cell preparation should result in the appearance of newly synthesized, "heavy" receptors that can be resolved from old, "light" receptors using techniques that resolve proteins based on their relative buoyant density. For acetylcholine receptors, old (light) receptors could be resolved from new (heavy) receptors using velocity sedimentation in sucrose-D_2O and sucrose-H_2O gradients, whereas isopycnic centrifugation in CsCl gradients has been used to resolve old from new receptors for insulin (Krupp and Lane [1982]) and EGF (Krupp et al. [1982]).

An important control for these experiments is to ensure that introduction of heavy amino acids into the cell culture medium does not change the receptor binding or receptor-mediated functional properties of the target cell. In addition, the investigator must develop a buoyant-density centrifugation protocol that sufficiently resolves old from new receptors such that quantitation of the area under the peaks for the sedimentation profile of the two receptor populations can be achieved. When this has been accomplished, and a parameter of the sedimentation profile (such as peak height or peak area) has been shown to reflect accurately the receptor concentration in the light or heavy receptor fraction, then the rate of receptor synthesis can be quantitated in a straightforward manner.

For determination of the rate of receptor synthesis using the heavy amino acid approach, cells are switched to heavy amino acid-containing medium at time 0. At time 0 and various times thereafter, cells are cooled to 4°C.

Radioligand binding to the cell surface then is performed at 4°C to prevent further turnover of the receptor population. At the end of the incubation, cells are washed with cold buffer to remove unbound radioligand and extracted with a suitable biological detergent that permits subsequent detection of ligand-receptor complexes. Alternatively, receptors can be extracted by detergent without prior occupancy with radioligand, in which case binding to unoccupied receptors can be determined following the centrifugation procedure. The 100,000 × g supernatant of the detergent extract is applied to an appropriate density gradient for resolving new (heavy) from old (light) receptors. The relative proportions of these two entities are then determined. Using these data, a progress curve can be plotted for the rate of heavy receptor synthesis and light receptor decay (cf. figure 6-5).

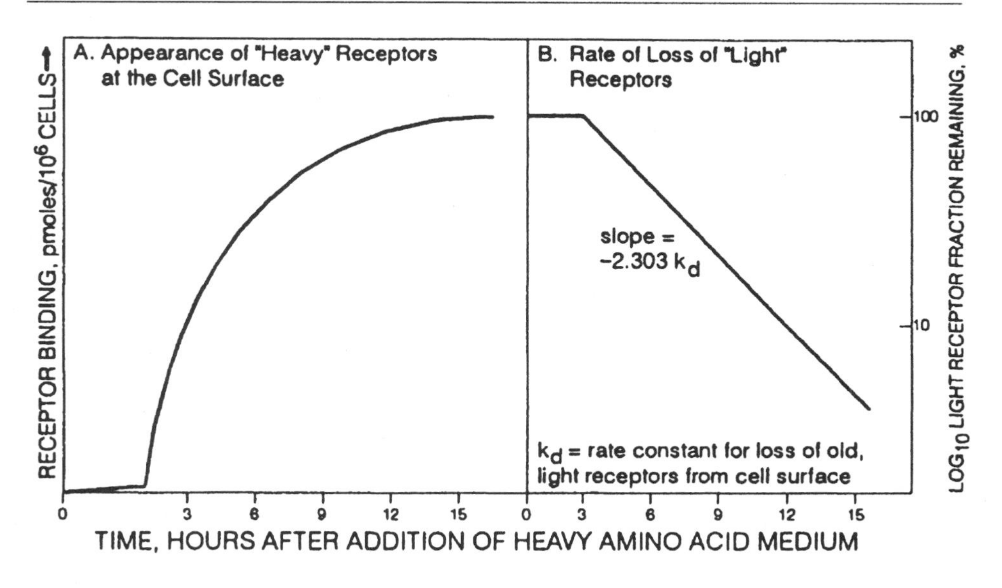

Figure 6-5. Determination of the rate constants of receptor turnover using the heavy amino acid "density shift" methodological approach.

The synthetic rate constant k_s can be determined from the limiting initial slopes of heavy receptor synthesis at time 0, and is expressed in units of pmol of receptor/cell number/unit of time. The first-order decay constant for receptors (k_d) can be determined by replotting the data for light receptor decay on a semi-log plot (In bound versus time) of the percentage of the "light" receptor population remaining. The slope of the line equals -k_d (or -2.303 k_d, if the data are plotted as Log_{10} [light receptor] versus time,. as in Figure 6-5b).

The heavy amino acid density shift technique is useful for determining several aspects of receptor turnover. For example, the delay between the time when culture medium is changed to contain heavy isotope-labeled amino acids and the time of heavy receptor appearance on the cell surface reflects the time of translation and processing of new receptors prior to insertion into the surface membrane. The quantity of light receptors that continue to appear at the cell surface following switching of the media to heavy amino acids may reflect the size of a precursor pool of receptors that is in transit between protein translation and cell surface expression or a pool of receptors that recycles from the surface membrane. In addition, the ability to assess both receptor synthesis (i.e., rate of appearance of heavy receptors) and receptor degradation (i.e., rate of disappearance of light receptors) makes this approach useful for assessing the contribution that varying rates of receptor synthesis versus degradation play in receptor down-regulation and up-regulation.

SUMMARY

The availability of morphological and biochemical methodologies to monitor receptor turnover permits investigators to answer a variety of questions regarding the relationship between the cellular distribution of a receptor population and receptor-mediated functions. Quantifying receptor turnover allows the impact of ligand or regulatory processes on receptor distribution to be assessed. As our understanding of the details of receptor trafficking increases, so will the importance of quantifying regulated steps in the overall life cycle of receptors.

REFERENCES

General

Brown, M.S., Anderson, G.W. and Goldstein, J.L. (1983) Recycling receptors: The round trip itinerary of migrant membrane proteins. Cell 32:663-667.

Geuze, H.J., Slot, J.W., Strous, G.J. A.M., Lodish, H.F. and Schwartz, A.L. (1983) Intracellular site of asialoglycoprotein receptor-ligand uncoupling: Double-label immunoelectron microscopy during receptor-mediated endocytosis. Cell 32:277-287.

Lauffenburger, D.A. and Linderman, J. (1993) Receptor/Ligand Trafficking. In *Models for Binding, Trafficking and Signaling*, pp. 73-132. Oxford UK: Oxford University Press.

Marchese, A., Chen, C., Kim, Y-M. and Benovic, J.L. (2003) The ins and outs of G protein-coupled receptor trafficking. Trends Biochem. Sci. 28(7):369-376.

Pearse, B.M. and Robinson, M.S. (1990) Clathrin, adaptors, and sorting. Ann. Rev. Cell Biol. 6:151-171.

Sorkin, A. and von Zastrow, M. (2002) Signal Transduction and Endocytosis: Close Encounters of Many Kinds. Nat. Mol. Cell. Biol. 3:600-614.

Turvy, D.N. and Blum, J.S. (1998) Detection of biotinylated cell surface receptors and MHC molecules in a capture ELISA: a rapid assay to measure endocytosis. M. Immunol. Meth. 212:9-18.

Wiley, H.S. and Cunningham, D.D. (1982) The endocytotic rate constant: A cellular parameter for quantitating receptor-mediated endocytosis. J. Biol. Chem. 257:4222-4229.

Wiley, H.S. (1992) Receptors: Topology, dynamics and regulation. In *Fundamentals of Medical Cell Biology*, Vol. 5A, "Membrane Dynamics and Signalling," pp. 113-142. JAI Press, Inc.

Specific

Barak, L.S., Tiberi, M., Freedman, N.J., Kwatra, M.M., Lefkowitz, R.J. and Caron, M.G. (1994) A Highly Conserved Tyrosine Residue in G Protein-coupled Receptors Is Required for Agonist-mediated β_2-Adrenergic Receptor Sequestration. J. Biol. Chem. 269(4):2790-2795.

Berry, S.A., Shah, M.C., Khan, N. and Roth, B.L. (1996) Rapid Agonist-Induced Internalization of the 5-Hydroxytryptamine$_{2A}$ Receptor Occurs via the Endosome Pathway *In Vitro*. Mol. Pharmacol. 50:306-313.

Bhatnagar, A., Willins, D.L., Gray, J.A., Woods, J., Benovic, J.L. and Roth, B.L. (2001) The Dynamin-dependent, Arrestin-independent Internalization of 5-Hydroxytryptamine 2A (5-HT_{2A}) Serotonin Receptors Reveals Differential Sorting of Arrestins and 5-HT_{2A} Receptors during Endocytosis. J. Biol. Chem. 276:8269-8277.

Carpenter, G. and Cohen, S. (1976) ^{125}I-labelled human epidermal growth factor: Binding, internalization and degradation. J. Cell Biol. 71:159-171.

Daunt, D.A., Hurt, C., Hein, L., Kallio, J., Feng, F. and Kobilka, B.K. (1997) Subtype-Specific Intracellular Trafficking of α_2-Adrenergic Receptors. Mol. Pharm. 51:711-720.

Devreotes, P.N., Gardner, J.M. and Fambrough, D.M. (1977) Kinetics of biosynthesis of acetylcholine receptor and subsequent incorporation into plasma membrane of cultured chick skeletal muscle. Cell 10:365-373.

Edson, M.G. and Liggett, S.B. (1992) Subtype-selective desensitization of alpha$_2$- adrenergic receptors. J. Biol Chem. 267:25473-25479.

Fehlmann, M., Carpentier, J.-L., van Obberghen, E., Freychet, P., Thamm, P., Saunders, D., Brandenburg, D. and Orci, L. (1982) Internalized insulin receptors are recycled to the cell surface in rat hepatocytes. Proc. Natl. Acad. Sci. USA 79:5921-5925.

French, A.R., Sudlow, G.P., Wiley, H.S. and Lauffenburger, D.A. (1994) Postendocytotic trafficking of epidermal growth factor-receptor complexes is mediated through saturable and specific endosomal interactions. J. Biol. Chem. 269:15749-15755.

Goldstein, J.L., Basu, S.K., Brunschede, G.Y. and Brown, M.S. (1976) Release of low density lipoprotein from its cell surface receptor by sulfated glycosaminoglycans. Cell 7:85-95.

Goldstein, J.L., Anderson, R.G.W. and Brown, M.D. (1979) Coated pits, coated vesicles, and receptor-mediated endocytosis. Nature 279:679-685.

Haigler, H.T., Maxfield, F.R., Willingham, M.C. and Pastan, I. (1980) Dansylcadaverine inhibits internalization of ^{125}I-epidermal growth factor in Balb/c 3T3 cells. J. Biol. Chem. 255:1239-1241.

Herbst, J.J., Opresko, L.K., Walsh, B.J., Lauffenburger, D.A. and Wiley, H.S. (1994) Regulation of postendocytotic trafficking of the epidermal growth factor through endosomal retention. J. Biol. Chem. 269:12365-12373.

Hertel, C., Coulter, S.J., Perkins, J.P. (1985) A comparison of catecholamine-induced internalization of β-adrenergic receptors and receptor-mediated endocytosis of epidermal growth factor in human astrocytoma cells. Inhibition by phenylarsine oxide. J. Biol. Chem. 260:12547-12553.

Hilairet, S., Belanger, C., Bertrand, J., Laperriere, A., Foord, S.M. and Bouvier, M. (2001) Agonist-promoted Internalization of a Ternary Complex between Calcitonin Receptor-like Receptor, Receptor Activity-modifying Protein 1 (RAMP1), and β-Arrestin. J. Biol. Chem. 276(45):42182-42190.

Insel, P., Mahan, L.C., Motulsky, H.J., Stoolman, L.M. and Koachman, A.M. (1983) Time-dependent decreases in binding affinity of agonists for β-adrenergic receptors of intact S49 lymphoma cells. A mechanism of desensitization. J. Biol. Chem. 258:13597-13605.

Jenkinson, K.M., Southwell, B.R. and Furness, J.B. (1999) Two affinities for a single antagonist at the neuronal NK_1 tachykinin receptor: evidence from quantitation of receptor endocytosis. Br. J. Pharm. 126:131-136.

Kadle, R., Kalter, V.G., Raizada, M.K. and Fellows, R.E. (1983) Cycloheximide causes accumulation of insulin receptors at the cell surface of cultured fibroblasts. J. Biol. Chem. 258:13116-13119.

Kallal, L., Gagnon, A.W., Penn, R.B. and Benovic, J.L. (1998) Visualization of Agonist-induced Sequestration and Down-regulation of a Green Fluorescent protein-tagged β_2-Adrenergic Receptor. J. Biol. Chem. 273(1):322-328.

Krupp, M.N., Connolly, D.T. and Lane, M.D. (1982) Synthesis, turnover and downregulation of epidermal growth factor receptors in human A431 epidermoid carcinoma cells and skin fibroblasts. J. Biol. Chem. 257:11489-11496.

Krupp, M.N. and Lane, M.D. (1982) Evidence for two different pathways for the degradation of insulin and insulin receptor in the chick liver cell. J. Biol. Chem. 257:1372-1377.

Lloyd, C.E. and Ascoli, M. (1983) On the mechanisms involved in the regulation of the cell surface receptors for human choriogonadotropin and mouse epidermal growth factor in cultured Leydig tumor cells. J. Cell Biol. 96:521-526.

Parker, E.M., Swigart, P., Nunnally, M.H., Perkins, J.P. and Ross, E.M. (1995) Carboxyl-terminal domains in the avian β_1-adrenergic receptor that regulate agonist-promoted endocytosis. J. Biol. Chem. 270(12):1-6.

Picus, R.C., Shreve, P.C., Toews, M.L., Bylund, D.B. (1993) Down-regulation of alpha$_2$-adrenoceptor subtypes. Eur. J. Pharm. 244:181-185.

Pittman, R.N. and Molinoff, P.B. (1980) Interactions of agonists and antagonists with β-adrenergic receptors in intact L6 muscle cells. J. Cycl. Nucl. Res. 6:421-435.

Ronnett, G.V., Tennekoon, G., Knutson, V.P. and Lane, M.D. (1983) Kinetics of insulin receptor transit to and removal from the plasma membrane. J. Biol. Chem. 258:283-290.

Schwartz, A.L., Fridovich, S.E. and Lodish, H.F. (1982) Kinetics of internalization and recycling of the asialoglycoprotein receptor in a hepatoma cell line. J. Biol. Chem. 257:4230-4237.

Simantov, R. and Sachs, L. (1973) Regulation of acetylcholine receptors in relation to acetylcholinesterase in neuroblastoma cells. Proc. Natl. Acad. Sci. USA 70:2902-2905.

Snavely, M.D., Ziegler, M.G. and Insel, P.A. (1985) A new approach to determine rates of receptor appearance and disappearance *in vivo*. Mol. Pharmacol. 27:19-26.

Sneddon, W.B., Syme, C.A., Bisellow, A., Magyar, C.E., Rochdi, M.D., Parent, J-L., Weinman, E.J., Abou-Samra, A.B. and Friedman, P.A. (2003) Activation-independent Parathyroid Hormone Receptor Internalization Is Regulated by NHERF1 (EBP50). J. Biol. Chem. 278(44):43787-43796.

Staehelin, M. and Simons, P. (1982) Rapid and reversible disappearance of β-adrenergic cell surface receptors. Eur. Mol. Biol. Org. (EMBO) J. 1:187-190.

Staehelin, M. and Hertel, C. (1983) [^{3}H] CGP-12177, a β-adrenergic ligand suitable for measuring cell surface receptors. J. Recept. Res. 3:35-43.

Staehelin, M., Simons, P., Jaeggi, K. and Wigger, N. (1983) CGP-12177. A hydrophilic β-adrenergic receptor radioligand reveals high affinity binding of agonists to intact cells. J. Biol. Chem. 258:3496-3502.

Starbuck, C. and Lauffenburger, D.A. (1992) Mathematical model for the effects of epidermal growth factor receptor trafficking dynamics in fibroblast proliferation responses. Biotechnol. Prog. 8:132-143.

Stoscheck, C.M. and Carpenter, G. (1984) Down-regulation of epidermal growth factor receptors: Direct demonstration of receptor degradation in human fibroblasts. J. Cell. Biol. 98:1048-1053.

Tanowitz, M. and von Zastrow, M. (2003) A Novel Endocytic Recycling Signal That Distinguishes the Membrane Trafficking of Naturally Occurring Opioid Receptors. J. Biol. Chem. 278(46):45978-45986.

Toews, M.L. and Perkins, J.P. (1984) Agonist-induced changes in β-adrenergic receptors on intact cells. J. Biol. Chem. 259:2227-2235.

Toews, M.L., Waldo, G.L., Harden, T.K. and Perkins, J.P. (1984) Relationship between an altered membrane form and a low affinity form of the β-adrenergic receptor occurring during catecholamine-induced desensitization: Evidence for receptor internalization. J. Biol. Chem. 259:11844-11850.

Vieira, A. (1998) ELISA-based assay for scatchard analysis of ligand-receptor interactions. Mol. Biotechnol. 10(3):247-250.

Waters, C.M., Oberg, K.C., Carpenter, G. and Overholser, K.A. (1990) Rate constants for binding, dissociation, and internalization of EGF: effect of receptor occupancy and ligand concentration. Biochemistry 29(14):3563-3569.

INDEX